Environmental Footprints and Eco-design of Products and Processes

Series editor

Subramanian Senthilkannan Muthu, SgT Group and API,
Hong Kong, Hong Kong

This series aims to broadly cover all the aspects related to environmental assessment of products, development of environmental and ecological indicators and eco-design of various products and processes. Below are the areas fall under the aims and scope of this series, but not limited to: Environmental Life Cycle Assessment; Social Life Cycle Assessment; Organizational and Product Carbon Footprints; Ecological, Energy and Water Footprints; Life cycle costing; Environmental and sustainable indicators; Environmental impact assessment methods and tools; Eco-design (sustainable design) aspects and tools; Biodegradation studies; Recycling; Solid waste management; Environmental and social audits; Green Purchasing and tools; Product environmental footprints; Environmental management standards and regulations; Eco-labels; Green Claims and green washing; Assessment of sustainability aspects.

More information about this series at http://www.springer.com/series/13340

Subramanian Senthilkannan Muthu
Editor

Energy Footprints of the Food and Textile Sectors

Editor
Subramanian Senthilkannan Muthu
SgT Group and API
Hong Kong, Kowloon, Hong Kong

ISSN 2345-7651   ISSN 2345-766X   (electronic)
Environmental Footprints and Eco-design of Products and Processes
ISBN 978-981-13-2955-5   ISBN 978-981-13-2956-2   (eBook)
https://doi.org/10.1007/978-981-13-2956-2

Library of Congress Control Number: 2018960179

© Springer Nature Singapore Pte Ltd. 2019
This work is subject to copyright. All rights are reserved by the Publisher, whether the whole or part of the material is concerned, specifically the rights of translation, reprinting, reuse of illustrations, recitation, broadcasting, reproduction on microfilms or in any other physical way, and transmission or information storage and retrieval, electronic adaptation, computer software, or by similar or dissimilar methodology now known or hereafter developed.
The use of general descriptive names, registered names, trademarks, service marks, etc. in this publication does not imply, even in the absence of a specific statement, that such names are exempt from the relevant protective laws and regulations and therefore free for general use.
The publisher, the authors and the editors are safe to assume that the advice and information in this book are believed to be true and accurate at the date of publication. Neither the publisher nor the authors or the editors give a warranty, express or implied, with respect to the material contained herein or for any errors or omissions that may have been made. The publisher remains neutral with regard to jurisdictional claims in published maps and institutional affiliations.

This Springer imprint is published by the registered company Springer Nature Singapore Pte Ltd.
The registered company address is: 152 Beach Road, #21-01/04 Gateway East, Singapore 189721, Singapore

This book is dedicated to:
The lotus feet of my beloved
Lord Pazhaniandavar
My beloved late Father
My beloved Mother
My beloved Wife Karpagam and
Daughters—Anu and Karthika
My beloved Brother
Everyone working in the food and Textiles
sectors to make it ENVIRONMENTALLY
SUSTAINABLE

Contents

Energy Footprints of Food Products 1
P. Senthil Kumar and A. Saravanan

Energy and Carbon Footprint of Food Industry 19
S. Naresh Kumar and Bidisha Chakabarti

Energy Footprints of Textile Products 45
P. Senthil Kumar and G. Janet Joshiba

Energy Footprints of Food Products

P. Senthil Kumar and A. Saravanan

Abstract In numerous regions, ecological issues that are both local (for instance, high rates of urbanization, mechanical exercises, arrive utilize changes, or rural practices,) and worldwide (for instance, desertification, or deforestation) have significantly lessened the capacity of land to ingest CO_2. A few endeavors have been made to orchestrate rules for natural footprints of food. The energy footprint, similar to the environmental footprint, is a marker of advance that can be utilized as methods for activating activity at the neighborhood level. It is an effectively comprehended idea as it scales the message down to the level of a person. It additionally legitimizes associations and cooperation among various partners to discover new, economical and less harming arrangements. The data exhibit that the impressions are critical, both locally, national and comprehensive and have comes about for overall sustenance security and condition prosperity and effectiveness. The writing about concurs that worldwide sustenance creation framework produces impressive natural impressions and the circumstance would likely get troubling.

Keywords Food waste · Green house emission · Sustainability
Natural footprints

1 Introduction

Energy utilize is relied upon to increment every year, the current pattern toward growing more residential vitality sources is driven to some degree by political shakiness in some oil-rich countries and to a limited extent by the want to keep up vitality security (Yergin 2006). Mechanical headways, for example, level boring in conjunction with pressure driven cracking have made extraction of shale assets financially reasonable,

---

P. Senthil Kumar (✉)
Department of Chemical Engineering, SSN College of Engineering, Chennai 603110, India
e-mail: senthilchem8582@gmail.com; senthilkumarp@ssn.edu.in

A. Saravanan
Department of Biotechnology, Rajalakshmi Engineering College, Chennai 602105, India

© Springer Nature Singapore Pte Ltd. 2019
S. S. Muthu (ed.), *Energy Footprints of the Food and Textile Sectors*, Environmental
Footprints and Eco-design of Products and Processes,
https://doi.org/10.1007/978-981-13-2956-2_1

advancing a quick increment in unpredictable vitality creation throughout the most recent span (Kerr 2010). Simultaneously, acknowledgment of the impending communal and natural implications of environmental revolution is lashing the drive to direct emanations by extending carbon impartial wellsprings of vitality, for example, sun oriented and airstream control (Pimentel et al. 2002).

Real changes in farming innovation, foundation, and cultivating administration rehearses are required at this point. Through deteriorating speculations and prevailing tasks, the plan for worldwide supportability in sustenance generation frameworks appear to be overwhelming. This session distils modules from antiquity and from prevailing examinations to pick the correct ventures, arrangements, and official buildings to guarantee that water and vitality assets are utilized carefully in the testing a long time ahead (Khan and Hanjra 2009). Environmental change and populace development may have huge ramifications for rural creation and its ecological impression, particularly for inundated farming which gives around 40% of worldwide sustenance creation from only 18% of cropland. Rainfed sustenance creation frameworks will likewise go under extreme weight because of movements in climate designs also, changes in precipitation occasions and hydrological administrations and more noteworthy reliance ashore and water assets, bringing on additional assets corruption and dissolving efficiency.

Every footprint imagines the stream of vitality (as fuel, power, or steam) to significant end utilizes as a part of assembling, including boilers, control generators, process warmers, process coolers, machine-driven hardware, office warming, ventilation, and aerating and cooling, and lighting. The impressions exhibit information at two levels of detail. The principal page gives an abnormal state perspective of essential vitality (offsite and on location), while the second page indicates subtle elements of how vitality is circulated to nearby end employments. Note that vitality expended as a feedstock (i.e., nonfuel vitality supply that is changed over to made item and not utilized for warmth, power, or power age) is excluded in the vitality esteems exhibited in impressions.

Total information gave in every one of the divisions incorporates:

- Power and steam created offsite and exchanged to the office, and in addition power and steam produced nearby
- Fuel, power, and steam devoured by significant end utilizes as a part of an assembling office
- Offsite and on location vitality misfortunes because of the age, transmission and dissemination, and end utilize utilization of vitality (a few misfortunes are unrecoverable)
- Ozone depleting substance outflows discharged amid the burning of fuel.

Energy Footprints of Food Products

1.1 Energy Footprint

The significance of the vitality impression free of the biological impression in vitality situation investigation by utilizing an input—yield examination (IOA) based casing work (Ferngs 2002). It was previously displayed as a sub-pointer of the organic impression, addressing the measure of timberland zone that is being prerequisite to hold CO_2 outflows from oil subordinate consuming and power age utilizing sequestration regards for a world-ordinary forest (Wackernagel and Rees 1996). Contingent upon refreshed information acquired from the Diplomatic Panelon Weather amendment, the estimation of the vitality impression has been reconsidered with a small amount of around 30% of the aggregate anthropogenic discharges for sea take-up (Borucke et al. 2013).

1.1.1 Segments and Units

The vitality impression can be arranged into solid segments, for example, the petroleum product impression, the water power impression, and the atomic impression (Browne et al. 2009), which are all communicated as the region of backwoods that is important to make up for human-incited carbon dioxide (Van Den Bergh and Verbruggen 1999). The unit of estimation can be nearby acreage with a particular decarbonization evaluate. Fundamental normal for the energy footprints was shown in Table 1. Rundown of methodologies connected to the investigations of vitality impressions from the writing was shown in Table 2.

Table 1 Fundamental normal for the energy footprints

S. No	Item	Energy footprint
1	Reasonable roots	Ecological footprint
2	Research stressors	Forest for absorbing—Energy-related green house gases outflows
3	Impression components	Fossil fuel, hydroelectricity, atomic, etc.
4	Metric units	Area-based (gha, ha, etc.)
5	Count methods	NFA, NFA-PLUM, IOA, base-up, top-down
6	Information availability	Medium
7	Methodological standardization	Low
8	Weighting accuracy	Not applicable
9	Resultant interpretation	High
10	Geographical specification	Medium
11	Worldwide comparability	High
12	General applicability	Low

Table 2 Rundown of methodologies connected to the investigations of vitality impressions from the writing

S. No	Question/scale	Energy footprint	References
1	Item/sustenance/energy	Bottom-up approach Top-down approach	Santek et al. (2010) Stoglehner (2003)
2	Association/segment/exchange/building	IOA Top-down approach NFA NFA-PLUM	Ferng (2002) Chen and Lin (2008) Browne et al. (2009) Wiedmann (2009)
3	Area/country/world	NFA	Chen and Lin (2008)

2 Obscure Issues

2.1 Designate Deforestation to Arrive Utilize Exercises

Deforestation is regularly connected with the extension of rural zone, in specific for domesticated animals touching and development of soybean and palm oil, yet additionally different products, for example, espresso, cocoa, banana. Be that as it may, there is no clear strategy for dispensing these weights to rural items (Hernandez et al. 2000; Ruf and Schroth 2004; Steinfeld et al. 2006; Fitzherbert et al. 2008; Flysjo et al. 2011; Ponsioen and Blonk 2012; Aide et al. 2013; Hylander et al. 2013; Hortenhuber et al. 2014). The explanation behind this is there is never an unmistakable direct connection between arrive change and agrarian exercises, dissimilar to for instance the connection between rural hardware or agrochemical inputs and agrarian exercises. The issue ought not to be roughly, in what way to remunerate weights of the previous to present terrestrial usage as projected at first in PAS 2050:2008 and still connected in numerous natural impressions thinks about. This alleged direct land utilize switch standard wholes up all the hurtful mediations of the arrive change movement and distributes them to every time of terrestrial usage in the initial 20 eons subsequently change. The thought originates from the avoidance time of the Inter-governmental Panel for Climate Change (IPCC) Guidelines in which dirt natural carbon swings starting with one harmony then onto the next after land change. Applying this idea for intercessions identified with soil and over-the-ground carbon in LCA includes a few issues: the horticultural creation in the initial 20 years can differ from zero (neglected) to low-income domesticated animals brushing or profoundly beneficial soybean generation. Much trickier is that after 20 a long time, the rural action is not related with weight on timberlands by any stretch of the imagination. This outcomes in an abnormal discretionary edge that as far as anyone knows isolates the devastating soya and oil palm from the sans disafforestation soya and oil palm.

2.2 Survey a Rural Emanations Profile

Agribusiness is additionally connected to emanations of different hurtful substances, for example, pesticides, smelling salts, nitrate, phosphorus, phosphate, overwhelming metals, nitrogen oxides, nitrous oxide, and methane. The greater part of these emanations is an aftereffect of organic procedures in the dirt, yield or creature framework. The issue is that they are greatly hard to quantify or to ascertain with any sensible accuracy. There are numerous conditions and definite estimation models accessible, all subsequent in various results (e.g. Kros et al. 2012). A conceivable outcome is that having more itemized figurings may bring about a bigger impression, which isn't inspiring organizations to put resources into extra endeavors for gathering the required information for these models. Then again, basic models neglect to precisely recognize the moderation possibilities of enhanced administration (Cederberg et al. 2013).

Straightforward outflow models require little information yet have a tendency to be erroneous, as they don't consider the intricate substances of natural and physical procedures in soils, products and creatures. Complex discharge models permit better thought of these procedures and of the impacts of agriculturist hones on them and on resulting discharges. Such models can create more exact outflow gauges and in this manner decrease vulnerabilities in discharge information (Dufosse et al. 2013). In any case, composite representations want additional information that may not generally be accessible. Display precision increments with show intricacy, however this connection is not direct, with expanding levels of multifaceted nature, unavoidable losses as far as intricacy are acquired (Robinson 2008). In this manner prescribe to use at any rate models of moderate multifaceted nature, gave that input information are accessible and that the model is adequately easy to use.

2.3 Pick the Utilitarian Unit of Nourishment and Refreshments

Other than the decent variety of natural effects, sustenance and drink items have a substantial range of functionalities. This makes it trying to think about the natural execution of sustenance and drink items. It additionally expresses that the practical unit characterizes the measurement of the distinguished capacities (execution qualities) of the item. Along these lines, satisfactorily choosing a useful unit is of prime significance, on the grounds that diverse useful unit could prompt distinctive outcomes for a similar item frameworks (Reap et al. 2008). Be that as it may, a precise technique for indicating practical units has yet not been created. A great many people concur that nourishment and drinks fill in as a wellspring of supplements, for example, protein, vitamins, minerals, filaments, and so on. Anyway there is no particular pointer that gets every single one of these nutritious constraints. Additionally, there are likewise separates in taste, surface, and additional particular assets.

Table 3 Cases of as often as possible utilized practical units, items they were utilized for and relating references

S. No	Functional unit	Products
1	Mass	Strong sustenance
2	Volume	Refreshments
3	Protein	Creature based items Meat replacers
4	Energy content	Sustenance
5	Serving	Sustenance
6	Wholesome file	Sustenance
7	Monetary esteem	Sustenance

As an outcome, the natural impressions of sustenance and drinks are hard to look at on a totally reasonable premise. One way to tackle this issue is to extend the two item frameworks for correlation with finish eats less with the same nourishing esteems. Nonetheless, this includes again a considerable measure of self-assertive presumptions that influence the outcomes. Additionally, the thought is to have natural impressions of individual items and not of weight control plans or ways of life. Other proposed functional units are giving nourishing worth (protein or nourishing record), a serving, a measure of vitality, and a monetary esteem was shown in Table 3.

2.4 Manage Multi-useful Procedures in Agri-Sustenance Frameworks

Multi-usefulness is extremely basic in agrarian and agro-mechanical procedures. A couple of cases are straw and grain, cotton and fiber, oilseed oil and devour, corn wet preparing co-things, deplete and animals for butcher or then again swelling, meat and butcher outcomes, and so forth. In such examples of multi-helpfulness, it must be picked how to divide data and yield surges of the method among the co-things. Following the ISO chain of importance, need to investigate the likelihood of framework development. This noteworthy approach is in principle conceivable by subtracting the natural weights of items that would have been created when no co-generation occurred. By and by, it is possible, for instance as performed by Dalgaard et al. (2008) for soybean oil and feast. In any case, this prompts logical inconsistency on the choices of substituted items. The substituted things never have the simple same properties. Another frustrating component is that the substituted things may in like manner be from multi-valuable methodology. For example, soybean oil may supplant pam oil, anyway then the palm parcel devour necessities to exchange for example sunflower supper, et cetera.

As per ISO, "Where assignment cannot be kept away from, the sources of info and yields of the framework ought to be divided between its diverse items or capacities in a way that mirrors the fundamental physical connections between them". A critical issue with this decide is that you initially need to choose which of the yields are really items and which yields are simply buildups that can be grabbed for nothing and utilized for moving up to important items. For instance, some side-effects, for example, straw, fertilizer, beet mash, brewer's grains, and so forth could in a few cases be named such buildups, however the definitions and contrasts between results and deposits are unclear. For example, the Renewable Energy Directive arranged "cultivating harvest stores, including straw, bagasse, husks, cobs and nutshells, and developments from taking care of, including grungy (glycerine that isn't refined)" to have zero life cycle loads, while these may have a market regard. The fundamental dependable way to deal with arrange this is to choose whether the yield has a market regard (co-or result) or not (development or misuse).

Once that is done, a physical normal for vital information, for example, the creature bolster for domesticated animals frameworks, or of the co-items is for the most part chosen. Largely, this is the vitality content, since vitality is frequently considered (by non-nutritionists) as the principle capacity of bolster or sustenance. In some cases the objective of an examination is to think about the natural effect of protein-rich nourishment. In those examinations, some of the time the decision is made to utilizes the protein content for dispensing the weights. In any case, the unmistakable verification of a run of the mill property, for example, imperativeness or protein content does less develop or mirror a normal association between co-things. Any two things on the planet have a mass and essentialness content; anyway, this does not mean they are associated. The connection between co-items is their regular beginning, along these lines, agreeing to ISO a physical designation technique ought not be founded on the physical properties of the co-items, yet Or maybe on the physical or organic system mirroring their regular cause, as this is the premise of their relationship.

3 Energy: A Global Challenge

Environmental change is an unpretentious type of human rights infringement. There is no immediate abuse or danger, however burning of non-renewable energy sources in industrialized countries has risked the capacity of specific social orders to keep up their conventional works on, lessening their social personality and their association with their indigenous habitat.

3.1 The Footprint of Different Activities

This measure can likewise be introduced as far as the sorts of items or administrations gave by the worldwide hectares, for instance, as far as merchandise from edit lands, creature items, angle, woodland items, developed regions, and vitality and water utilize. Such investigations recognize which zones are putting the best strains on biological systems, and can help set arrangement needs. Development in creature items and vitality utilize, particularly of petroleum derivatives, are two regions that are quickly expanding these strains.

3.2 The Footprint of Nations

Natural Footprint takes a gander at the aggregate sum of worldwide hectares that are required to help a specific populace, paying little heed to whether those hectares are inside the national fringes where that populace lives. It does this by thinking about the net utilization of the populace (or movement) of enthusiasm, subtracting the worldwide hectares utilized for send out from those utilized for imports and creation. Energy footprint: measures up was shown in Fig. 1.

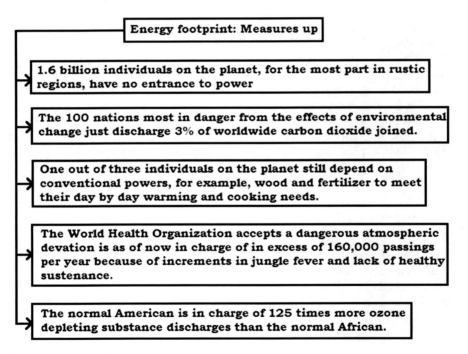

Fig. 1 Energy footprint: measures up

3.3 A Walkthrough of the Footprints

These impressions depend on an investigative model that was produced utilizing division particular vitality measurements, ignition related green house gas outflows information and rules, and the direction of industry specialists. Figure 1 demonstrates the shading legend utilized as a part of the footprints. Every impression comprises of two figures; the main figure offers a review of the area's aggregate essential vitality stream including offsite vitality and misfortunes, while the second figure shows a more point by point breakdown of the on location vitality stream. The expression "Add up to" in the impressions alludes to the aggregate whole of offsite and on location esteems. In vitality terms, this is alluded to as aggregate essential vitality. Footprint legend was shown in Fig. 2.

The footprint pathway catches both vitality free market activity. On the supply side, the impressions give points of interest on vitality buys and moves into a plant site (counting fills got from on location delivered results), and on location age of steam and power. On the request side, the impressions show the end utilization of vitality inside a given area, from process vitality uses, for example, warmers and engines to nonprocess uses, for example lighting. The impressions distinguish where vitality is lost because of age and appropriation misfortunes and framework wasteful aspects, both inside and outside the plant limit. Misfortunes are basic, as they speak

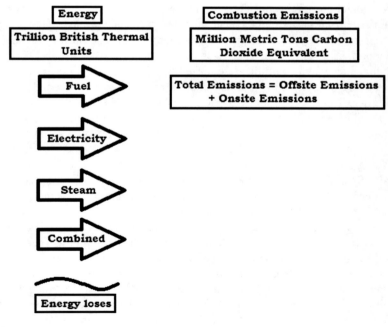

Fig. 2 Footprint legend

to potential chances to enhance effectiveness and decrease vitality utilization through best vitality administration practices and advancements.

Energy losses happen along the whole vitality pathway from age and conveyance to end utilize. Energy is lost in creating and transmitting force and steam and in process and nonprocess end utilization of energy, steam, and fuel. In the impression examination these vitality misfortune esteems are assessed in light of writing pursuit and companion survey. Since vitality misfortunes shift enormously by industry and by office, proper segment wide and part particular vitality misfortune gauge suspicions are made with the understanding that these evaluations are profoundly subject to the particular assembling plant site.

The energy footprints depend on real plant review information and in this manner speak to a sensible circulation of vitality utilize and misfortunes over the division overall. Through them, we can start to evaluate the extent of vitality utilization and misfortunes, both by end utilize and fuel write. They likewise give a gauge from which to ascertain the advantages of enhancing vitality productivity. The green house gas esteems in the impression can be utilized to help green house gas administration arranging and investigation.

4 Climate Chaos

Environmental change is considered by numerous researchers to be the most genuine risk confronting individuals and the earth on which we as a whole depend. The impacts of environmental change incorporate rising ocean levels, more serious tempests and surges, longer dry seasons and the spread of tropical sicknesses. Environmental switch could likewise make up to 200 million atmosphere exiles by 2050, driven from their homes because of surges, dry season or water and sustenance deficiencies. The poorer you are the less you have added to the issue of environmental change, yet the more you are probably going to endure. The world's poorest individuals have the most minimal vitality or carbon impression—they utilize the minimum vitality thus make the slightest ozone harming substances—yet are probably going to endure the most. Researchers foresee that in Africa in the vicinity of 75 and 250 million individuals may encounter genuine water deficiencies because of environmental change. Nourishment edit yields may decrease by as much as a half.

There is a reasonable connection between environmental change and human rights. The privilege to life itself can be denied in the quick result of an outrageous climate occasion, for example, a typhoon. More outrageous tempests are anticipated because of environmental change. Individuals' entitlement to nourishment may likewise be influenced. Sustenance generation is probably profitable to decrease because of expanded temperatures, variations in precipitation designs, soil disintegration, and desertification and rising ocean levels which will make seaside arrive unusable for developing products. As the earth gets hotter, warm waves and water deficiencies

Energy Footprints of Food Products 11

will make it hard to get to safe drinking water and sanitation. The outcome could be expanded clashes between nations over constrained water. Environmental change will impactsly affect the privilege to wellbeing. It will, for instance, give a superior atmosphere to intestinal sickness conveying mosquitoes.

5 Worldwide Farming Energy Footprint

5.1 Energy Factor

Energy resources are a basic element of worldwide monetary development and improvement (Chow et al. 2003). Energy utilization devises suggestions for monetary development; the neighborhood, countrywide, and worldwide condition, and notwithstanding for the worldwide amity and safety. Worldwide vitality utilization can be ordered into five noteworthy parts: business, conveyance, agribusiness, business and open administrations, and private. Agribusiness is a noteworthy maker of vitality also, creating a lot of vitality through photosynthesis action by harvests and manors. Sun powered vitality rules the vitality adjust on sustenance generation, representing 90% of aggregate vitality inputs even in concentrated frameworks.

Agricultural tasks influence a genuinely little commitment to the general vitality to utilize. For example, the utilization of homestead apparatus, water system, treatment and synthetic pesticides adds up to just 3.9% of the business vitality utilize. Of this, 70% is related with the creation and utilization of synthetic composts (Vlek et al. 2004). In any case, energy inputs have expanded lopsidedly after some time. A progression of cross-sectional ponders on cultivating frameworks propose that yield increase demonstrate declining comes back to vitality contributions after a specific point.

Balancing out the carbon dioxide initiated segment of anthropogenic environmental change is a vitality issue and a noteworthy pathway to decreasing the ecological impression of vitality utilizes. From farming point of view this incorporates ideal utilization of manure vitality; soil carbon sequestration ventures; and biofuel crops.

5.2 Agricultural Chemicals

Fertilizer is a vital circuitous vitality contribution for sustenance creation. By a wide margin nitrogen compost is the biggest compost vitality input. Nitrogen manures require around 10 times more imperativeness to convey per ton than phosphorous and potassium fertilizers, and they routinely represent 55–65% of on-develop essentialness use for exceptional yield crops. The rate of aggregation of soil natural issue is frequently upper on prepared pitches because of higher biomass generation, be that as it may, it conveys a carbon "cost" that is infrequently surveyed in the type of carbon discharges amid the generation and application of inorganic compost. Carbon bookkeeping utilizing an existence cycle approach can give a more educated perspective of such oversights (Stewart et al. 2005).

5.3 System for Situation Examinations of Energy Footprints

Final energy utilization is chiefly contributed to creation exercises that give merchandise and enterprises for family unit utilization. An expansion in the last interest for merchandise and administrations would subsequently bring about an expansion in last vitality utilization at a given level of innovation (Wackernagel et al. 1999). The required last utilization of energies can be upheld by various sorts of essential energies, every one of which has its individual vitality arrive proportions. These circumstances ought to be thought about unequivocally in the count of vitality impressions; however they were overlooked in the first technique. Figure 2 demonstrates four factors (household last request, essential vitality change framework, vitality arrive proportions, and neighborhood accessible vitality lands) and their linkages that are thought to be thought about explicitly in the calculation of vitality impressions. Proposed computation system for energy footprints was appeared in Fig. 3.

The local last request is portrayed here as the last demand recorded in a cutting edge trades table (or called an input– yield table) subtracting the admissions. The household last request subsequently involves family usage, government utilize, wander, and stock. This paper business the local last request to address the last request of the inhabitants living in a portrayed economy or nation. The family use can be viably secluded from the family unit last usage, if essential, by a couple of changes in the states of the propagation models.

One obscure authority proposes pointing out that imperativeness land for CO_2 sinks should be limited to the boondocks for whole deal maintenance of CO_2, and that business woodlands should not be consolidated in light of the fact that CO_2 is released again when wood things, for instance, papers and furniture are seared or spoiled. In any case, the neglected components and their linkages have an indispensable part to

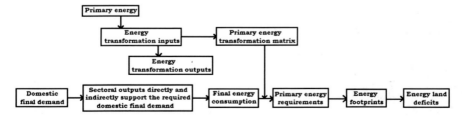

Fig. 3 Proposed computation system for energy footprints

Energy Footprints of Food Products

play in finding out the fundamental energies exemplified in the stock and endeavors ate up by a portrayed human people, and in coordinating circumstance examinations of the plan instruments for diminishing essentialness impressions.

5.4 Renovation Aspects of Food Footprint

The institutionalization methodology is the essential procedure for figuring food ecological footprint. In this examination, it is finished by changing for close-by yield factors remembering the ultimate objective to measure the advantages need unequivocally and particularly.

(i) Conversion variables of cropland for various food items

Many food items start from crops developed in the farmland like grain or soybeans, and optional items counting meat, drain, egg, and cultivated fish are bolstered by crops and results got from farmland, these all have diverse cropland prerequisites. When we change over these food items to developed territory, a simple path is to change over them into grain reciprocals by proportional variables. It's focused on that no land necessities were doled out to side-effects and squanders from the food industry that are utilized as a part of the foods of household creatures to keep away from twofold including. No land is relegated to creature fat in light of the fact that the proportional factor of meat has thought about the entire land assignment. For soya oil the proportional factor is chosen by the proportion of soybean-to grain-normal yield and change proficiency in the nourishment business. Soya oil creates a lot of oil cakes that are utilized for domesticated animals grains and no land is relegated. Cultivated fish is sustained by feed from cropland, and aquatic products by fish catch are computed by transformation factor of angling arrive.

(ii) Conversion variables of cropland and grassland

In some animals creation frameworks one-food items are delivered, while in different frameworks two nourishment things are delivered. Pig keeping brings about the creation of just pork, while dairy cultivating brings about crude milk and beef. At the point when more items are made in a generation framework, the aggregate land necessities were partitioned over the items corresponding to their vitality yield, similar to milk and meat. Another purpose behind the contrast between their change factors is the domesticated animals, for example, dairy livestock and lamb are encouraged by grassland, feed consumption from the rangeland. Conversion variables of cropland and grassland for mutton, milk and beef was shown in Table 4.

(iii) Transformation elements of national normal energy

Imperativeness drive is associated with account the exemplified essentialness creation and transportation of foodstuff, and it can be obtained from Redefining Progress. About more than 80% of the sustenance usage things are arranged and packaged

Table 4 Conversion variables of cropland and grassland

S. No	Consumption items	Mutton	Milk	Beef
1	Cropland (kg/kg)	2.3	0.3	3.6
2	Grassland(kg/hm^2)	24.3	216	18
3	Proportion from grazing	35%	26%	14%
4	Altered mean change components of prairie (kg/hm^2)	69.4	830.8	128.6

in made countries. So Energy expended in sustenance creation and transportation is ascertained by increasing the global vitality thickness with the coefficient 0.4 for improving the count. Vitality arrive speaks to the measure of woodland arrive expected to take-up anthropogenic carbon outflows. The adjusting carbon consistency standard of boondocks is settled considering the carbon pool, carbon dioxide movement of tropical mountain rain woodlands organic network and the genuine timberland gainfulness. Carbon power speaks to carbon outflows of various sort of fossil vitality. The change factor of household fossil vitality is ascertained via carbon force separated retention rate through afforestation deducting the portion consumed by sea.

6 Food Is Important to Our Ecological Footprint

Our selection of nourishments can possibly increment or diminishing the impacts of a worldwide temperature alteration. Issues, for example, sustaining the growing total populace and having enough nourishment and water are likewise part of the master plan about utilizing our assets carefully. By picking nourishment that has less bundling, has not voyage immense separations and has been delivered reasonably, we can help decrease our impression. The accompanying practices are related with being a mindful national as for biological manageability.

(a) Purchase sustenance that is privately delivered
The closer the homestead is, the less fuel is expected to transport the sustenance to the table. Largely, nourishment voyages long separations, requiring refrigeration and capacity. The general stores need to offer privately developed items.

(b) Buy from nearby agriculturists' market
An agriculturists' market is one in which nearby ranchers or makers offer their own particular deliver. All items sold ought to have been developed, raised, gotten, blended, salted, prepared, smoked or handled by the stallholder.

Energy Footprints of Food Products

(c) Plant a garden and develop claim new create
Developing your own products of the soil decreases the vitality and waste that regularly goes into getting nourishment from the field to your plate, for example, transport, refrigeration and bundling. You can likewise screen the sum and kind of manure and supplements used to develop your nourishment.

(d) Avoid handled sustenance
The make of prepared nourishments utilizes a lot of vitality, water and materials underway, chilling, bundling and transport, and creating a lot of waste. Purchase crisp products of the soil as opposed to canned or solidified (in spite of the fact that the last are in some cases a need in remote networks).

(e) Eat sustenance in season
On the off chance that the required foods grown from the ground are not accessible, pick those that are in season. On the off chance that you eat products of the soil out of season, recall that they have either voyage long separations from a place where they are in season as well as have utilized vitality in chilly stockpiling. Purchasing privately developed, occasional sustenance implies a decrease in nourishment miles, less vitality utilized as a part of capacity and less bundling required safeguarding new deliver.

(f) Buy natural nourishment
Natural and different types of low-input cultivating that utilization insignificant or no pesticides and composts—which are vitality serious in their produce—devour up to 40 for each penny less vitality, and bolster larger amounts of untamed life on ranches. Natural and other all-encompassing cultivating approaches have a tendency to organize creature welfare more than customary techniques. Purchasing sustenance that is more natural can decrease your nourishment impression by around 5 for every penny.

(g) Choose sustenance that have no or insignificant bundling
A lot of assets are required to create the bundling and bundle the item. The transfer of the bundling has additionally negative effects on nature. (viii) Drink tap water rather than filtered water. Filtered water costs around 500 times more than tap water, and the business emanates a great many huge amounts of CO_2 consistently. Drinking filtered water offers no demonstrated medical advantages. Since no materials are utilized as a part of its individual bundling.

(h) Eat less meat and eat more plant-based sustenance
Meat and dairy items are the most asset concentrated and the slightest fuel-productive sustenance we have. Substantial amounts of vitality are required to develop, gather, and ship creature feed; house, transport and butcher creatures; process and bundle their meat; and refrigerate it until the point when it's cooked. A solitary serve of meat is evaluated to make five kilograms of ozone harming substances. It takes 1350 L of water to deliver a kilogram of wheat however it takes 16,000 L of water to create a kilogram of hamburger.

(i) Buy manageable fish
More than seventy five percent of the world's angling grounds have been over-fished to the point that they are presently underneath manageable levels. You can utilize your buyer capacity to ensure imperiled species.

(j) Don't squander sustenance
   Purchase just what you will eat. Check the utilization by date on the item mark. The generation of sustenance utilizes water and vitality and the nourishment we don't eat makes squander. Numerous people and families could diminish utilization of sustenance by and large.
(k) Buy reasonably exchanged sustenance and drink
   When you purchase nourishment from abroad, endeavor to purchase decently exchanged items, for example, those confirmed by Fairtrade. These items support interest in individuals—advancing social equity, nearby monetary improvement and reasonable costs.

7 Reducing

In relationship with oil and gas (and each oil subordinate), wind imperativeness has the most insignificant lifecycle surges of carbon dioxide and other ozone hurting substances (Jacobson 2008). Numerous investigations have shown a critical misfortune in worldwide biodiversity and biological system benefits because of expanding worldwide temperatures from the utilization of petroleum products (McDaniel and Borton 2002). In that capacity, wind vitality advancement is being advanced as a "spotless" option. In any case, this point of view regularly neglects the consistently developing effects of vitality advancement on the scene, which have been named vitality sprawl (McDonald et al. 2009). Like oil and gas, wind vitality requires a system of streets, transmission lines, and related foundation to catch and transport the power.

The impacts of oil, gas, and wind imperativeness progression on various neighborhood and scene level pointers that may affect terrestrial biodiversity and natural framework organizations, including untamed life mortality, condition mishap, region crack, disturbance and light pollution, prominent species, and changes in carbon stocks and freshwater resources. These pointers were picked as surrogates for assessed impacts on species tolerable assortment and the game plan of some organic framework organizations, which are site specific and difficult to get. These markers are not by any stretch of the imagination likewise essential to biodiversity and condition organizations, and they are appeared in different courses, dependent upon imperativeness create.

8 Conclusion

The consequences of an ecological impression of a sustenance or refreshment item very rely upon the decisions made in regards to the five issues examined here. Unmistakably announcing these viewpoints, completing an affectability examination and watchful elucidation are along these lines vital for achieving the correct conclusions.

Energy Footprints of Food Products

In spite of the fact that any suggestion comes about because of a bargain or the absence of better option, our proposals as spurred in this paper are, in outline, to: utilize the strategy to figure the measure of land utilize change caused by expanded weight from developing regions per edit in every nation; dependably ascertain the farming discharges utilizing no less than a halfway level of detail, gave that info information are accessible, with the alternative to include more detail; set up regionalized water shortage evaluations in view of good quality water system and yield evapotranspiration information, and at any rate nation particular water shortage factors; report the ecological impression comes about per unit of monetary esteem other than per unit of mass or volume to empower more pleasant correlations; and apply monetary designation in all multi-useful rural and agri-modern procedures in light of sensible costs figured as various year midpoints.

References

Aide TM, Clark ML, Grau HR et al (2013) Deforestation and reforestation of Latin America and the Caribbean (2001–2010). Biotropica 46:262–271

Borucke M, Moore D, Cranston G et al (2013) Accounting for demand and supply of the biosphere's regenerative capacity: the national footprint accounts' underlying methodology and framework. Ecol Indic 24:518–533

Browne D, O'Regan B, Moles R (2009) Use of ecological footprinting to explorealternative domestic energy and electricity policy scenarios in an Irish city-region. Energ Policy 37:2205–2213

Cederberg C, Henriksson M, Berglund et al (2013) An LCA researcher's wish list—data and emission models needed to improve LCA studies of animal production. Animal 7:212–219

Chen CZ, Lin ZS (2008) Multiple timescale analysis and factor analysis of energyecological footprint growth in China 1953–2006. Energ Policy 36:1666–1678

Chow J, Kopp RJ, Portney PR (2003) Energy resources and global development. Science 302:1528–1531

Dalgaard R, Schmidt J, Halberg N, Christensen P, Thrane M, Pengue A (2008) LCA of Soybean Meal. Int J LCA 13:240–254

Dufosse K, Gabrielle B, Drouet JL et al (2013) Using agroecosystem modelling to improve the estimates of N_2O emissions in the life cycle assessment of biofuels. Waste Biomass Valor 4:593–606

Ferng JJ (2002) Toward a scenario analysis framework for energy footprints. Ecol Econ 40:53–69

Fitzherbert EB, Struebig MJ, Morel A et al (2008) How will oil palm expansion affect biodiversity? Trends Ecol Evolut 23(10):538–545

Flysjo A, Cederberg C, Henriksson M et al (2011) How does co-product handling affect the carbon footprint of milk? Case study of milk production in New Zealand and Sweden. J Life Cycle Assess 16:420–430

Hernandez C, Witter SG, Hall CA et al (2000) The Costa Rican banana industry: can it be sustainable. Quantifying Sustain Dev: The Future of Trop Econ 563–593

Hortenhuber S, Piringer G, Zollitsch W et al (2014) Land use and land use change in agricultural life cycle assessments and carbon footprints-the case for regionally specific land use change versus other methods. J Clean Prod 73:31–39

Hylander K, Nemomissa S, Delrue J et al (2013) Effects of coffee management on deforestation rates and forest integrity. Conserv Biol 27(5):1031–1040

Jacobson MZ (2008) Review of solutions to global warming, air pollution, and energy security. Energy Environ Sci 2:148–173

Kerr RA (2010) Natural gas from shale bursts onto the scene. Science 328:1624–1626

Khan S, Hanjra MA (2009) Footprints of water and energy inputs in food production—global perspectives. Food Policy 34:130–140

Kros H, Lesschen JP, Bannink A et al (2012) Review on calculation methodologies and improvement of mechanistic modelling of GHG emissions. DLO, Wageningen, Netherlands

McDaniel CN, Borton DN (2002) Increased human energy use causes biological diversity loss and undermines prospects for sustainability. Bioscience 52:929–936

McDonald RI, Fargione J, Kiesecker J et al (2009) Energy sprawl or energy efficiency: climate policy on natural habitat for the United States of America. PLoS ONE 4:6802. https://doi.org/10.1371/journal.pone.0006802

Pimentel D, Herz M, Glickstein M et al (2002) Renewable energy: current and potential issues. Bioscience 52:1111–1120

Ponsioen TC, Blonk TJ (2012) Calculating land use change in carbon footprints of agricultural products as an impact of current land use. J Clean Prod 28:120–126

Reap J, Roman F, Duncan S et al (2008) A survey of unresolved problems in life cycle assessment. Int J LCA 13:290–300

Robinson S (2008) Conceptual modeling for simulation Part I: definition and requirements. J Oper Res Soc 59(3):278–290

Ruf F, Schroth G (2004) Chocolate forests and monocultures: a historical review of cocoa growing and its conflicting role in tropical deforestation and forest conservation. In: Agro forestry and biodiversity conservation in tropical landscapes. Island Press, Washington, pp 107–134

Santek B, Gwehenberger G, Santek MI et al (2010) Evaluation of energy demand and the sustainability of different bioethanol production processes from sugar beet. Resour Conserv Recycl 54:872–877

Steinfeld H, Gerber P, Wassenaar T et al (2006) Livestock's long shadow. FAO, Rome, p 229

Stewart WM, Dibb DW, Johnston AE et al (2005) The contribution of commercial fertilizer nutrients to food production. Agron J 97(1):1–6

Stoglehner G (2003) Ecological footprint—a tool for assessing sustainable energysupplies. J Clean Prod 11:267–277

Van den Bergh JCJM, Verbruggen H (1999) Spatial sustainability, tradeand indicators: an evaluation of the 'ecological footprint'. Ecol Econ 29:61–72

Vlek P, Rodraguez-Kuhl G, Sommer R (2004) Energy use and CO_2 production in tropical agriculture and means and strategies for reduction or mitigation. Environ Dev Sustain 6(1):213–233

Wackernagel M, Lewan L, Hansson CB (1999) Evaluating the use of natural capital with the ecological footprint. Ambio 28:604–612

Wackernagel M, Rees WE (1996) Our ecological footprint: reducing human impact on the earth. New Society, Gabriola Island, British Columbia

Wiedmann T (2009) A first empirical comparison of energy footprints embodiedin trade—MRIO versus PLUM. Ecol Econ 68:1975–1990

Yergin D (2006) Ensuring energy security. Foreign Aff 85:69–82

Energy and Carbon Footprint of Food Industry

S. Naresh Kumar and Bidisha Chakabarti

Abstract Food processing is a major thriving industry globally and provides livelihood to millions of workers. Food processing is an energy intensive process and often has an impact on the environment which remains undiagnosed and hence not quantified. Food processing industry comprises the organized as well as unorganized sector with varying levels of energy requirement and therefore the carbon foot prints also significantly vary. Higher energy use is often related to higher greenhouse gas (GHG) emission which is responsible for global warming and climate change. Carbon footprint (CFP) of food industry is an estimate of the energy use and GHG emissions caused due to the processing and delivery of food items to the consumer and also disposal of packaging. Recently there is a growing interest in estimating the carbon footprint of food industries to know how improved technologies can be used to make food processing less energy and carbon intensive. In this book chapter we would like to provide an overview of energy use and carbon footprint of different types of food industries. Quantification of CFP is generally done using Life Cycle Assessment (LCA) in which GHG emissions are measured from the very beginning of the production process to its final use and disposal. GHG emission from a food industry will include both direct emissions as well as indirect emissions. The CFP of different sectors like fruit and beverage industry, sugar production, dairy sector, fisheries, meat and poultry supply chains are presented. Apart from this, research gaps and possible steps to minimize the carbon footprint will be mentioned. Assessing the CFP of food industries can help in identifying the GHG sources and can be useful in developing alternative technologies which are more energy efficient and reduces GHG emission. Further, change in dietary pattern also contributes immensely to reduce the environmental impact of food consumption.

Keywords Food chain · Beverages · Dairy · Poultry · Fisheries · Sugar · Energy Carbon footprint

---

S. Naresh Kumar (✉) · B. Chakabarti
Centre for Environment Science and Climate Resilient Agriculture, ICAR-Indian Agricultural Research Institute, New Delhi 110012, India
e-mail: nareshkumar@iari.res.in; nareshkumar.soora@gmail.com

© Springer Nature Singapore Pte Ltd. 2019
S. S. Muthu (ed.), *Energy Footprints of the Food and Textile Sectors*, Environmental Footprints and Eco-design of Products and Processes,
https://doi.org/10.1007/978-981-13-2956-2_2

1 Introduction

In the 21st century mankind is facing three challenges i.e. food security, climate change, and energy security (Lal 2010). Over the past few decades, various footprint indicators have been introduced to raise awareness about the anthropogenic pressure exerted on the environment (Fang et al. 2014). The footprint concept originated from the idea of ecological footprint introduced in the 1990s (Rees 1992; Wackernagel and Rees 1996). The energy footprint which is a component of ecological footprint is defined as the area required to sequester the CO_2 emission arising from burning of fossil fuel, buffering the nuclear power radiation, and for building dams to generate hydro-electricity (Wackernagel et al. 1999). Many times, energy footprint, comprises a dominant portion of the ecological footprint at regional or global level (Kitzes et al. 2009). Energy footprint values are represented in global hectare or local hectare with a specific carbon sequestration estimate (Walsh et al. 2010). Components of energy footprint include fossil fuel footprint, hydroelectricity footprint and nuclear power footprint (Čuček et al. 2012; Stöglehner 2003). The importance of energy efficiency was highlighted by Intergovernmental Panel on Climate Change (IPCC) highlighting the necessity to improve energy use efficiency for reducing greenhouse gas (GHG) emission (Metz et al. 2001).

Energy consumption in food production is largely at four levels viz., agriculture, transportation, food processing and food handling. Agriculture which deals with production of raw material for food processing consumes energy that is used for food production. Energy is also consumed for transportation and food processing. This indicates that the energy use efficiency needs to be improved in handling and transportation steps.

Efficient energy use is important for sustainable agricultural production since it helps in financial savings, preserves renewable fossil fuel, and reduces air pollution (Esengun et al. 2007). Major energy consumption can be classified into five sectors i.e. industry, agriculture, transportation, residential, commercial and public services (Khan and Hanjra 2009). Agricultural production is vulnerable to changes in availability of energy inputs. On farm activities during crop production uses around 2–5% of the commercial energy in almost all countries (World Bank 2007). Among this agricultural operations contribute to a small amount while the major part is associated with manufacture and use of chemical fertilizers (Vlek et al. 2004). Quantification of energy footprint can help in better understanding the use and distribution of energy and to compare its use and losses in different operations. Sectors with higher energy consumption and inefficient energy uses provide opportunities for improving the efficiency by developing new technologies or implementing better management practices. Developing systems with low energy input for a unit of product output will help in reducing emission of greenhouse gases in agricultural sector (Dalgaard 2000).

Increasing demand of readymade meals has increased the degree of processing in the food industry leading to higher energy use (Sonesson et al. 2004). Some steps of food processing consume more energy than others. For example, energy consumption in a fruit and vegetable plant is significantly higher for 50% for refrigeration and cooling followed by 40% for machine drives and 5% each for HVAC and lighting (Masanet et al. 2008). This energy requirement is tracked through the carbon foot print (CFP) as carbon is the surrogate for transfer of energy in a system. Carbon footprint of food gives an estimate of the greenhouse gas (GHG) emission caused from manufacture till delivery of the particular food item. The CFP is total set of greenhouse gas (GHG) emission caused by a product and is expressed in terms of carbon dioxide (CO_2) equivalent. Quantification of CFP is generally done using Life Cycle Assessment (LCA) also known as 'cradle to grave' analysis in which GHG emissions are measured from the start of the production process till its final use and disposal. Food chain activities such as pre-production, production, transport, storage, cooking and wastage of food are contributors to GHG emission (IPCC 2007; Garnett 2008; Chakrabarti et al. 2015). The food processing plants in Sweden used 5.75 TWh fuel and electricity (Statistics Sweden 2002). Among this total energy use, 45% was from electricity and 55% purchased fuels not including fuel for transports (Fritzson and Berntsson 2006).

Product carbon footprint (PCF) is a tool to quantify greenhouse gas (GHG) emissions from a product over entire supply chains, starting from procurement of raw material, production, processing, value addition, packaging, storage, transport, use, cooking, food waste and disposal.CFP is a quantitative expression of GHG emissions which helps in emission management and evaluation of mitigation practices (Carbon Trust 2007). Carbon footprints of products can help in identifying the GHG emission 'hotspots' in the processing and delivery activities, and guide manufacturers to identify ways to save energy (Bokern 2010).Besides providing information about GHG emission, carbon audits can help companies to evaluate their operations and making them more energy and emission efficient. Companies, policy makers and other stakeholders use CFP values to gain an understanding on the GHG emissions from various products. Recent studies have indicated that food and drink, transportation, and construction sectors are the most significant contributors of GHG emissions (UNEP 2008).

Emission from food items comprises three main GHGs like carbon dioxide (CO_2), methane (CH_4), and nitrous oxide (N_2O) (Kling and Hough 2010). Carbon dioxide is released when fossil fuels are burnt to generate energy. Methane is emitted from paddy fields, livestock digestion and during decomposition of food waste in landfills while nitrous oxide comes out from application of fertilizers for growing crops. According to Kling and Hough (2010) food represents 21% of the total annual carbon footprint of an average American. In UK, food is found to be responsible for approximately one fifth of all GHG emissions (Berners-Lee et al. 2012; Garnett 2008). In Asia, food and drink, transportation, and construction sectors contribute between 70–80% of the continent's total GHG footprint (UNEP 2008). In USA packaging of food accounts for 10% of all food production emissions with cartons and aluminum packaging being the major contributors (Kling and Hough 2010), though buying fresh foods can help in lowering the CFP.

The methodologies for quantification of carbon footprint of products are still evolving (Pandey et al. 2011). Quantification of CFP is done by taking all major GHGs into account. Comparison among different non-CO_2 GHGs are done by converting their effect to the common unit of carbon dioxide "equivalent" (CO_2eq) based on their global warming potential, relative to that of CO_2 (ITC 2012). Methodologies and standards for GHG accounting are given by IPCC 2006 guidelines, GHG protocol of World Resource Institute (WRI), ISO 14064 (parts 1 and 2), Publicly Available Specifications-2050 (PAS 2050) of British Standard Institution (BSI), ISO 14025, ISO 14067 (Pandey et al. 2011). According to IPCC guidelines, GHG emission is calculated by Tier 1, Tier 2 or Tier 3 approach. Moving up to higher tiers improves the accuracy of GHG assessment and reduces the uncertainty (Chakrabarti et al. 2015).

GHG emission from a food industry will include both direct emissions as well as indirect emissions. Direct energy use is for on-farm activities at raw-material production and during various steps of manufacturing processes, while indirect energy use is during storage, transportation and use of electricity for running the food industry. As mentioned earlier, this later part of food processing, transport and storage consume more energy. Nowadays there is an increase in readymade meals which means an increase in food processing industries. This leads use of more energy (Sonesson et al. 2004). All these cause more emission of GHG. Intergovernmental Panel on Climate Change (IPCC) has developed a standard methodology to quantify GHG emissions from major economic sectors (IPCC 2000, 2006), which is used by most of the countries to calculate GHG emissions. This chapter gives an overview of the energy foot print and CFP of different food industries, elaborates how carbon intensive different food items are and also gives an idea about the challenges of assessing energy foot print (EFP) and CFP of different food sectors.

2 Food Miles

Production and distribution of food items has been known to be a major source of GHG emission. Food-miles are a measure of how far the food travels from its production point to the point of the final consumer (Christopher and Matthews 2008). Globalisation of the food and distribution industry as well as demand for seasonal goods throughout the year, and concentration of sales in supermarkets have led to increasing food miles (Padfield et al. 2012). Reducing the distance a food travels before it reaches the consumer can help in lowering the energy requirement and CFP as well as will have a positive effect on the climate. But it is not always true that local foods are less carbon intensive. If producing food locally emits more carbon than transporting them then it can be counter-productive (Kling and Hough 2010). Transport modes like ocean and rail will be more carbon efficient on per-weight basis, even over long distances (Wakeland et al. 2012). A comparison of emission intensities of different transport modes indicated that found that air freighting is more emission intensive than others modes, like water and rail (Weber and Matthews 2008) (Table 1). But transport through water and rail are dependent upon the availability of navigable water and established railway tracks (Wakeland et al. 2012).

Energy and Carbon Footprint of Food Industry

It has been reported that transportation represents only 11% of total GHG emissions of food in USand red meat being around 150% more GHG-intensive than chicken or fish (Weber and Matthews 2008).

The agri-food industry consumes energy at various processes from crop production, transport, storage, processing and marketing. A typical energy input in some of the agri-food industries indicated fare distribution of energy input across sugar, oil, meat, beverages and frozen food as well as canned food products (Fig. 1).

However, estimating Gross Energy Requirement (GER), the sum of all non-renewable energy resources consumed in making available a product or service is important for improving energy use efficiency. The GER is expressed in energy units per physical unit of product or service delivered.

3 Sugar Industry

Production and processing of sugar is energy intensive process (Mendoza et al. 2004). In the sugar making activities, the main sources of GHG emissions are farming, raw

Table 1 Emissions from different modes of transport (*Source* Weber and Matthews 2008)

Transport mode	kg CO_2 eq per ton-km
International water	0.14
Inland water	0.21
Rail	0.18
Truck	1.8
Air	6.8

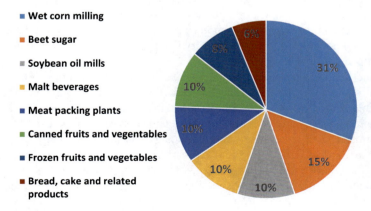

Fig. 1 Shares of agri-food industry (*Source* Okos et al. 1998)

material processing and transport (Klenk et al. 2012). In Thailand, quantification of CFP of sugar during the sugarcane cultivation and milling process showed that sugar production has a carbon footprint of 550 g CO_2 eq kg^{-1} sugar (Yuttitham et al. 2011). Out of this 490 g CO_2 eq kg^{-1} sugar was attributed to sugarcane cultivation and 60 g CO_2 eq kg^{-1} sugar to the milling process. The carbon footprint of raw sugar varied between 217 to 809 g CO_2 eq/kg sugar, while that of refined sugar varied between 329 to 1121 g CO_2 eq/kg (Fisher 2013). Assessment of GHG emission from sugar production in Brazil showed that 241 kg of CO_2 equivalent is released to the atmosphere per ton of sugar produced (de Figueiredo et al. 2010). Residue burning contributed to 44% of the total emission while 20% resulted from the use of fertilizers and 18% from burning of fossil fuel. According to this study reduction in CFP could be achieved by harvesting of sugarcane without burning of residues.

Haciseferogullari et al (2003) assessed the energy value of sugar beet grown in Turkey and found that energy value was 16.80 MJ kg^{-1}. Fertilizer had the largest share in total energy input followed by fuel and oil. Among total energy used in fuel and oil, 54.6% was contributed by fuel used for irrigation purpose share. It was suggested to optimize irrigation methods and method of fertilizer application to reduce the energy input. Rational use of water using solar irrigation pumps is one major opportunity to reduce load on fossil fuel based energy. However, it is to caution that irrigation water must be used judiciously, particularly with solar punp systems as over minig of ground water will cause undesirable effects such as ground water depletion and ecosystem imbalance. For sugarcane production, the energy requirement was estimated to be 0.43 GJ/one tonne cane while energy requirement for cane milling it was estimated at 1.95 GJ/one tonne cane (Mendoza and Samson 2002). The energy consumption for production of one litre of ethanol produced from molasses consumed 0.74–0.96 kWh.

Studies on product carbon footprint (PCF) of beet sugar indicated that PCF value was 610 g CO_2 eq kg^{-1} sugar for German sugar (Setzer 2005), while its was 1040 g CO_2 eq kg^{-1} sugar for US beet sugar (Fereday et al. 2010). On the other hand, the CFP of European Union beet sugar ranged from 242 to 771 kg CO_2 eq t^{-1} sugar (Klenk et al. 2012). Around 31.6% of these emissions are from sugar beet cultivation and 3.6% related to sugar beet transport (Fig. 2). The remaining is related to different processing operations in the sugar factory like production of steam, drying of sugar beet pulp and lime kiln operation.

Reduction in energy consumption in sugar industry can be done by implementing new equipment with modern technology (Pathak 1999). Sugarcane bagasse is a renewable source of energy which can be used as raw material for different value added products. By products like filtercake and molasses can be used for alcohol production. Use of solar energy and bio energy can improve energy efficiency of this sector (Pathak 1999). In South African the sugar industry produces sugar as well as electricity and also raw materials for ethanol production (Mbohwa 2013). This cogeneration of electricity was linked to energy security and reduced GHG emission. According to Mbohwa (2013), primary energy sources used for sugar processing were electrical energy and thermal energy. Technological improvement along

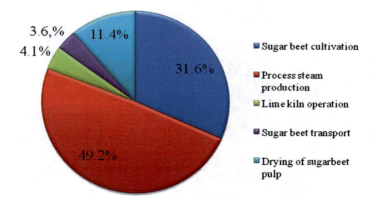

Fig. 2 The GHG emission from sugar beet growing, transport and processing (*Source* Klenk et al. 2012)

with improvements in process designing could improve energy efficiency in sugar factories.

4 Beverage Industry

Milk, tea, coffee, juice, beer and wine make up 28% of a consumer's total beverage consumption (Greenwich 2010). The beverage industry is very energy intensive. In beverage industry refrigeration make up the major portion of energy use while in confectionery, boilers and heat distribution systems make the largest contribution (Carbon Trust 2012). This industry is diverse with many sub-sectors, each employing a range of industrial processes. This sector is the largest energy user for refrigeration technology. Other energy intensive techniques include use of compressed air, use of electric motors, boilers, cooking, distillation, drying and evaporation (Carbon Trust 2012). Like in other food based industries, in the whole life cycle of a beverage energy input and GHG emission occurs during production phase at farm level, manufacturing and packaging phase in industries, transportation, in consumer home and from wastes generated from beverages. A typical energy input in a fruit juice industry indicates that generating steam consumes bulk of the energy (Fig. 3).

To address several issues of beverage industry, Beverage Industry Environmental Roundtable (BIER) was formed in August 2006. Companies like PepsiCo, Nestlé Waters, Coca-Cola Company are part of BIER. Research efforts were initiated in Europe and North America by BIER to determine the major contributors of GHG in the life cycle of carbonated soft drink (BIER 2010). The aim was to reduce energy consumption across the entire value chain, including agricultural practices, transportation, packaging and storage.

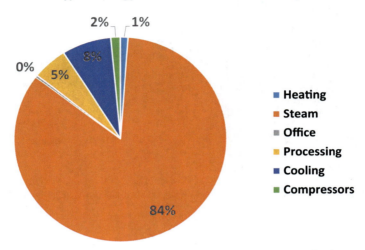

Fig. 3 Typical energy use in the fruit juice industry (*Source* Feld and Poremski 2016)

The food and drink processing industry is the fourth largest industrial energy user in UK. In year 2010, energy consumption in this sector was 37 TWh and emission was around 11 million tonnes of CO_2 into the atmosphere (Carbon Trust 2012). A study on UK beverage industry revealed that manufacturing and packaging were the main contributors to global warming (Amienyo and Azapagic 2011). Aluminium carbonated beverages contributed 80% of carbon emissions while beer carbon emissions were 51%. Analysis of the CFP of the Finnish beverage industry for the period of 2000 to 2012 indicated that the footprint of packaging materials increased while that of direct CO_2 emissions, energy use and waste material got decreased during this period (Karjalainen 2013). An assessment of the environmental impact of packaging and disposal of the Spanish juice, beer and water showed that recycling was the most environmentally friendly disposal option compared to incineration. With regards to the packaging options, the lowest environmental impacts were with aseptic carton and plastic packaging (Pasqualino et al. 2010). Saxe (2010) reviewed literature to summarize the environmental impact of beverages and concluded that 1 litre of beer caused CFP ranging from 1 to 1.5 kg CO_2 eq L^{-1} beer while for wine it ranged from 1.9 to 5.3 kg CO_2 eq L^{-1}.

Nutrient Density to Climate Impact (NDCI) index is an index which represents the nutrient density of each beverage combined with the GHG emissions to produce it (Smedman et al. 2010). The highest NDCI index values indicates the highest nutrient density scores in relation to the GHG emissions. Among the beverages like milk, orange juice, soft drink, wine, beer, bottled carbonated water, oat drink and soy drink, the NDCI was substantially higher (0.54) for milk than other beverages due to a high nutrient density (Smedman et al. 2010).

5 Dairy Industry

Milk production is an agricultural activity which has environmental side effects like greenhouse gas emission and nutrient enrichment in water bodies (Van Calker 2005). The dairy industry uses energy in several forms to produce different types of milk products. Dairy sector is becoming more energy intensive which results in more environmental and economic loads (Todde et al. 2018). In recent time, cost of milk and milk based products is rising due to the cost of raw materials and high energy required for dairy processing (Prabhakar et al. 2015). Dairy industry uses high amount of energy in processing, manufacture and storage of various products (Janzekovic et al. 2009). The industry handles large amount of milk and uses energy in as steam, hot water, chilled water compressed air, and electricity (Kalla et al. 2017). Maintaining high product quality and reducing cost of production can be met in this sector through investments in energy-efficient technologies. Different stages which require energy are milk reception and storage, filtration and standardization, homogenization and pasteurization. Electricity is used to drive motors, fans, pumps and compressed air systems, as well as for building and lighting systems (Kalla et al. 2017). Intensification of the dairy farms is associated with higher energy demand and associated increase in CO_2 emission. In 2007, the global dairy sector emitted 1969 Mt CO_2 equivalent of GHG of which 1328 Mt are attributed to milk production (FAO 2010a, b). The contribution of dairy sector to the total global anthropogenic GHG emissions was 4%.

The UK dairy industry produces over 13 billion litres of liquid milk each year and provides variety of dairy products. Energy performance varies across dairies with most efficient one using 32 kWh m^{-3} raw milk (Carbon Trust 2007) (Fig. 4). Difference in energy performance arises due to energy intensive evaporation and spray drying process carried out by some sites and also due to the variation in technologies adopted. According to Kalla et al. (2017) focused and strategic energy management programme can help in identifying and implementing energy efficient measures in the dairy industry. As per Australian dairy manufacturing environmental sustainability report of 2011, energy use in dairy was close to neutral with a 2% reduction in total energy but with a 3% increase for energy intensity of an additional 41 GJ per mL milk processed.

The dairy sector consumed 52, 34, 16 and 14 PJ primary energy in France, Germany, the Netherlands and United Kingdom in year 2000 (Ramirez et al. 2006). Figure 5 shows energy distribution in different processes involved in fluid milk processing and storage in Dutch dairies.

Meul et al. (2007) studied the changes in energy use efficiency in Flanders between periods 1989–1990 and 2000–2001 for dairy and arable farms and between 1989–1990 and 1997–1998 for pig farms. Mineral fertilisers and animal feed accounted for the maximum share of total energy use while use of diesel had the major share in direct energy use. The study reported that, total energy use per ha in dairy and arable farms decreased significantly over the considered time period

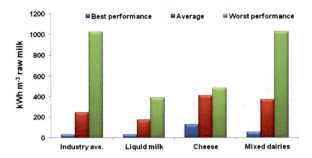

Fig. 4 Energy performance of dairies in UK (*Source* Carbon Trust 2008)

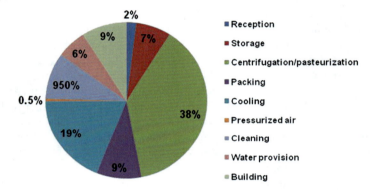

Fig. 5 Percentage of energy demand for different processes in Dutch dairies (*Source* Ramirez et al. 2006)

while in pig farms, energy use per fattening pig equivalent (FPE) in 1997–1998 was comparable to that in 1989–1990.

The GHG emissions from dairy sector include emissions from farm enteric fermentation, production of cattle feed and manure handling. Transport of milk to the processing farm, pasteurisation of milk, packaging, refrigeration, distribution and disposal of waste also contribute to the emissions. Methane is produced and emitted during enteric fermentation and manure management. An estimat-ed 137 Mt of CH_4 was produced in U.S. in 2011due to en-teric fermentation, which is 70% of the total agricultural methane emissions (Smith 2014). Enteric fermentation in livestock caused emission of 212.1Mt of CO_2 equivalent which constituted 63.4% of the total GHG emissions from agriculture sector in India (INCCA 2010). A detailed GHG emission inventory from Indian livestock indicated that dairy buffalo and indigenous dairy cattle contribute to 60% of the methane emissions and the per capita emission is 24.23 kg CH_4/animal/year (Chhabra et al. 2013).

The CFP of Canadian dairy industry was estimated using an integrated cradle to gate model (Vergé et al. 2013). Estimates of CFP for most of the dairy products ranged between 1 to 3 kg CO_2 eq kg^{-1} of product. Calculation of CFP in terms

Energy and Carbon Footprint of Food Industry

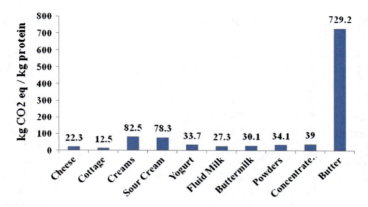

Fig. 6 Carbon footprint (CFP) of Canadian dairy products (*Source* Vergé et al. 2013)

of GHG emissions per kilogram of protein showed that cream and sour cream had higher CFPs with 83 and 78 kg of CO_2eq kg^{-1} of protein, respectively (Fig. 6). On the other hand, highest CFP was for butter, with 730 kg of CO_2eq kg^{-1} of protein.

Life cycle analysis of cheese production in USA showed that carbon footprint of cheddar cheese was 8.60 kg CO_2 eq kg^{-1} while for mozzarella cheese the value was 7.28 kg CO_2 eq kg^{-1} (Kim et al. 2013). Uzal (2013) analysed the energy use efficiency of two specialized dairy farms in Konya, Turkey. The study indicated animal feed contributed to a high percentage of total energy use while electricity accounted for the major portion of direct energy used. Energy productivity of the farms was 5.4 L milk per 100 MJ and 3.9 L milk per 100 MJ of energy use in free-stall dairy cattle housing system, and loose housing system respectively. Life cycle analysis of conventional and organic milk production in Netherlands revealed that energy use was higher in conventional farms than organic farms (Thomassen et al. 2008). This higher use of energy was due to more indirect energy use in conventional system as compared to organic system.

6 Meat and Egg Industry

Worldwide meat industry is a growing sector contributing to around 65% of global food production (FAOSTAT 2013). In meat processing facilities, use of electricity is the most dominating one (Canning et al. 2010). The energy use per one tonne of meat product has been increasing by about 14–48% in European Countries (Ramírez et al. 2006). Improving energy efficiency in industrial processes will reduce its dependence on fossil fuels and also lower down costs and GHG emission. Meat production is one of the main sources of GHG emission having potential impact on climate change. The share of animal agriculture to total global GHG emissions range between 10 and 25% (Steinfeld et al. 2006; Gill et al. 2010; Barclay 2012). According to the

Food and Agriculture Organization, world's livestock sector is responsible for 18% of the global GHG emission. Emissions from livestock constitute nearly 80 per cent of all agricultural emissions (Steinfeld et al. 2006; McMichael et al. 2007). Consumption of meat and dairy products was responsible for 14% of the total global warming within the 27 European Union member countries (Weidema et al. 2008). Beef industry uses lot of fossil energy (Tremblay 2000), most of which is used by feedlots where production operations are most concentrated and intensive. Energy consumption by the meat sector was 39, 35, 10 and 32 PJ in France, Germany, Netherlands and UK respectively in year 2001. Of this amount, between 40% and 60% was used by the further processing of meat sector (Ramirez et al. 2006). Fossil fuel is used for process heat while electricity is required for cooling. Feliciano et al. (2014) assessed the energy use efficiency in Portuguese slaughterhouses and found that electricity is mostly used energy source with about 50% used for meat processing and refrigeration. Munguia et al (2016) reported the results of an energy audit in a meat processing industry in north-western Mexico. Refrigeration is a key element in meat processing to ensure the meat quality. In order to achieve this lot of energy is spent in this sector. They found that ventilation and air conditioning processes had significant correlations in terms of energy saving. Illumination was second highest in terms of energy consumption and has improving potential as well.

Livestock buildings and poultry facilities are energy consumers. In the poultry sector energy is consumed for internal climate adjustment and operation of production equipment. A new technique to estimate the metabolisable energy (ME) requirement of broiler breeders for maintenance, bodyweight gain and egg mass and their respective energy efficiencies was developed using mathematical modelling. It is an accounting tool used for predicting the nutritional requirements for poultry with different genetic strains, environments and stages of meat gain or egg production. In modeling the daily ME requirement, ME requirement is partitioned into maintenance, egg mass and body weight gain. Using such approach, the estimated ME required for maintenance for breeders at 21 °C was 98.3 kcal per kilo bodyweight, 5.6 kcal per gram gain and 2.4 kcal per gram egg mass. The energy efficiencies for protein gain, fat gain and egg calories were 34, 79, and 65.7 per cent, respectively (Reyes et al. 2011). Technified poultry farms require more energy for providing thermal comfort, proper illumination levels, food dispensing and automated ventilation facility (Oviedo-Rondon 2009). According to international literature, energy consumption varies between 12–16 MJ/bird depending on the location of the poultry farm and the technology level used (Baxevanou et al. 2014). Baxevanou et al (2017) found significant variations in energy use among mountainous and lowland farms. In lowland farms 50% of the energy used was electrical while in mountainous farms it was only 20%. Higher demand for electricity in lowland farms was due to increased cooling needs. Insulation was suggested as an energy saving technique for mountainous farms. Pig farming also involves energy use for feeding, building services and waste removal. Some of the energy management options suggested by Carbon Trust (2005) are monitoring energy use, improvement in building insulation, carrying out maintenance and repair, use of enclosed creeps, using efficient lighting, fans and ducts and use of high efficiency motors for feed and waste handling.

Vergé et al. (2008) quantified the GHG emission from Canadian beef industry and found that total GHG emission from beef production have increased from 25 to 32 Tg of CO_2 equiv. between 1981 and 2001. But the emission intensity decreased from 16.4 to 10.4 kg of CO_2 eq kg^{-1} live animal weight during this period. Replacement of high roughage diet by high energy feeds led to reduction in enteric CH_4 emissions. Vries and Boer (2010) compared the environmental impact of production of beef, chicken, pork, egg and milk using life cycle analysis (LCA). LCA results showed that production of 1 kg beef used most land and energy, and had highest GWP, followed by pork, chicken, egg and milk (Fig. 7).

Arroyo-Pitacua et al (2015), assessed the energy requirement in a poultry farm in Central Mexico and found that energy requirement and CO_2 emission was maximum due to usage of LP gas in the heating system (Table 2). Results showed that to produce 1000 chickens CO_2 emission was 1206.4 kg CO2 eq. and energy productivity of the full production cycle was 0.12 kg MJ^{-1} (Table 3). In 2010 the global pig population was estimated at 968 million animals while the global poultry population was 22 billion animals with chickens making up 90% (FAOSTAT 2012). Energy use in pig farming is for building, animal feeding and waste removal.

Globally, GHG emissions from pig and chicken supply chains are estimated at 668 and 606 million tones of CO_2-equivalent, respectively (MacLeod et al. 2013). Although the most important source of CH_4 is the enteric fermentation from livestock but methane contribution from the pig and poultry industries originates from manures and from feed production. Carbon dioxide is emitted from on-farm energy use like cooling, ventilation and heating and also during transport of live animals and products to slaughter house and processing plants. Emission of N_2O occurs from manure management. Among the major livestock animals reared, GHG emission from poultry is the low-est (Dunkley and Dunkley 2013). Major-ity of the emissions from beef cattle (86%) and swine (69%) occurs at produc-tion stage while for poultry, 48% of the emissions were observed during the production stage and 52% from post production stage (Fig. 8).

In a recent analysis on energy conversion efficiency of US feed to food protein flux, it has been indicated that 31 Mt of protein from concentrates of corn, soybean, other crop s and by products; 21 Mt protein from processed roughage like hay, hay large, grass, silage and green chop and corn and sorghum silage; and 11 Mt protein from pasture was converted to animal products containing 4.7 Mt protein. This amounts to just about 8% conversion efficiency on weight basis. The efficiency varied among

Fig. 7 Global warming potential for livestock products, expressed in CO_2eq kg^{-1} of product (*Source* Vries and Boer 2010)

Table 2 Energy consumption in meat industry (adopted from Ramírez et al. 2006)

	Electricity	%	Fuels	%
	Activity		Activity	
Pork slaughtering	Cooling	40–70	Gas oven	60–65
	Slaughtering	5–30	Cleaning and disinfecting	18–20
	Water Cleaning	5–7	Singeing	15
	Lighting	2–8		
	Evisceration	3	Space heating	7
Cattle slaughtering	Slaughtering	26	Cleaning and disinfecting	80–90
	Evisceration	3		
	Cooling	45–70		
	Compressed air, lighteing and machines	30	Space heating	10–20
Poultry slaughtering	Cooling	52–60	Singeing	60
	Machines and compressed air	30	Cleaning and disinfecting	30
	Lighting and ventilation	4	Space heating	10
Meat processing	Cutting and mixing	40	Cleaning and disinfecting	25
	Cooling	40	Space heating	15
	Packing	10		
	Lighting	10		
Rendering	Compressed air, lighteing and machines	12	Vacuum evaporation	2
	Grinding and pressing	17	Drying	61
	Drying	23	Grinding and pressing	17
	Vacuum evaporation	6	Space heating	1
	Milling plant	8	Fat treatment	3
	Meal strelization	2	Meal sterilization	8

Table 3 Energy requirement and associated carbon dioxide (CO_2) emission during chicken production (1000 chickens)

Category	MJ	CO_2 eq.
Water pumping and distribution	33.4	4.2
Heating	22834	1156.6
Lighting	17.8	2.2
Food distribution	5.4	2.4
Ventilation	79	9.9
Hot air extraction	385	32.7
Total	23354.6	1206.4

Source Arroyo-Pitacua et al. (2015)

the type of animal protein that is produced. It was 9% for pork, 21% for poultry, 31% for eggs, 14% for dairy and 3% for beef protein (Shepon et al. 2016). In terms of energy, concentrates, processes roughage and pasture contributed 590, 316 and 280 Pcal yielding 12 Pcal of pork, 17 Pcal of poultry, 4 Pcal of eggs, 29 Pcal of dairy and 21 Pcal of beef (Shepon et al. 2016). This indicated that from 1187 Pcal of feed, 83 Pcal of edible animal products are produced at a conversion efficiency of just about 7% indicating a huge loss of 93% of energy.

7 Seafood Industry

Globally seafood is one of the most traded food products with around 40% of all fisheries and aquaculture production traded internationally (FAOSTAT 2012). Energy inputs in fishing activities include both direct and indirect. Direct energy inputs are required for propelling fishing vessels as well as deploying fishing gears. The three dominant forms of energy used in this include animate, wind, and fossil fuel energy.

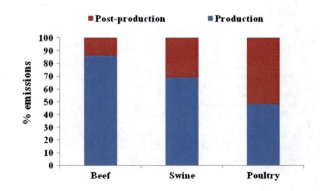

Fig. 8 Production and post-production emissions for beef swine and poultry (*Source* Dunkley and Dunkley 2013)

Indirect energy inputs are associated with building and maintenance of fishing vessels and providing fishing gear, bait, and ice (Tyedmers 2004).

The GER in fish harvesting up to the point of landing in the small-mechanised and non-mechanized sectors of Indian fisheries indicated that the GER t. fish^{-1} ranged from 5.54 and 5.91 GJ, respectively, for wooden and steel purse seiners powered by 156 hp engines; 6.40 GJ for wooden purse seiner with 235 hp engine; 25.18 GJ for mechanised gillnet/line fishing vessel with 89 hp engines; to 31.40 and 36.97 GJ, respectively, for wooden and steel trawlers powered by 99-106 hp engines (Boopendranath and Shahul Hameed 2013). The global analysis indicated that the energy use of various cultured products at farm gate varied from 3 MJ. Kilo per live weight for Galician mussels (Mytilus galloprovincialis) and Blue mussels (Mytilus edulis) to 13 for Striped catfish (Pangasianodon hypophthalmus); 26–98 for Atlantic salmon (Salmo salar); 62 for White-leg shrimp (Litopenaeus vannamei), 291 for Turbot (Scophtalmus maximus and 353 for Arctic char (Salmo alpinus). The variations in energy requirement is due to differences in methods of harvest and energy estimation (Hornborg and Ziegler 2014) (Table 4).

Recently, climate change is recognized as one of the critical issues for fisheries sector (Vivekanandan 2011). Use of fossil fuels for activities like fish catch, onboard processing and freezing of fishes cause considerable emissions of greenhouse gases. Compiling and analyzing fuel consumption, catch, and vessel/gear characteristic data from a wide range of sources, Tyedmers et al. (2005) concluded that maximum fuel is expended in near shore fishing grounds of the Northern Hemisphere. Fishing grounds with heavy fuel usage included the western Pacific and adjacent seas, the Bering Sea, and coastal waters of the north-eastern and south-western Atlantic and northern Indian Ocean. The variability in fuel use among fisheries depends on two factors i.e. the relative abundance and catchability of targeted species as well as the type of fishing gear employed (Tyedmers 2004). The fuel use intensity in cod

Table 4 Estimated energy use per unit protein energy output

Species group	Production types	Main species	Energy use per protein energy output (J/J)
Finfish, inland aquaculture	Ponds at different intensities	Carps	1–25
Finfish, mariculture	Marine cages	Atlantic salmon	40–50
Molluscs	Long line, tanks	Clams and cockles; oysters; mussels	10–136
Crustaceans	Ponds,	White leg shrimp	40–480
Other species		Amphibians and reptiles	
Algae		Kappaphycus alvarezii; Eucheuma spp.; Laminaria japonica	

Adopted from S Hornborg and F Ziegler https://swemarc.gu.se/digitalAssets/1536/1536133_ publication—energy-use-in-aquaculture.pdf accessed on 7th May, 2018; *Source* FAO (2012, 2014) and Troell et al. (2004)

fisheries where Danish seine nets were used was 440 L t^{-1} of landed fish while long line fisheries and trawl fisheries for cod had fuel use of 490 L t^{-1} and 530 L t^{-1} respectively (Tyedmers 2004). The carbon footprint of Tuna fisheries in Philippines is dominated by fuel combustion during fishing operations (Tan and Culaba 2009).The carbon footprint at the wholesale gate was estimated at 250–300 g CO_2 eq kg^{-1} of fish. But at the retail level it increased to 800–900 kg CO_2 eq kg^{-1}. The additional carbon footprint is incurred due to handling, storage and transport activities. The second most significant contributor to GHG emissions after fuel oil is refrigerant. A key refrigerant in use is R22, a hydro-chlorofluorocarbon (HCFC) with high ozone depletion and global warming potential of 1,810 (FAO 2012). Replacement of R22 is the most important potential improvement in the fishing phase (Winther et al. 2009).

An analysis on the environmental impact of fresh and canned mussel products indicated that purification stage had major impact in case of fresh mussels (Iribarren et al. 2010). The other significant contribution to global warming potential (GWP) was from transport to retailers. For canned mussels, the main source of GHG emissions to the environment was fuel oil production associated with ancillary operations followed by electricity production at various stages, and from tin can production. The quantification of partial carbon footprints of different activities involved in Tuna fisheries indicated that CFPs for fishing are much larger than other activities. The footprint of cooking, canning and transportation are of comparable magnitude, while that of cold storage is significantly small (Tan and Culaba 2009) (Table 5).

Technological improvements can help in lowering the CFP of fisheries sector. Implementing fuel efficiency norms for fishing vessels and shifting from fuel-intensive active fishing methods such as trawling to passive methods such as seining, lining and gillnetting may be useful in the long-term (Vivekanandan et al. 2013). Informing about the location of potential fishing zones to the fishermen has significantly reduced time and fuel energy required for marine fishing activity in some parts of India (CMFRI 2014), This also led to reduction in GHG emission from marine fishing activity.

Table 5 Partial carbon footprints of key processes in Tuna fisheries (Tan and Culaba 2009)

Process/activity	CFP (kg CO_2 eq kg^{-1} fish)
Fishing (Purse Seine)	1.15–5.27
Fishing (Long Line)	6.64–8.86
Fishing (Large Pump Boat)	2.11–4.70
Fishing (Small Pump Boat)	3.26–4.25
Cold Storage	0.0025–0.12
Cooking and Canning	0.42–1.38
Transportation	0.11–1.53

8 Dietary Pattern

There is considerable variation in GHG emissions among different food groups, with animal-based products having higher CFP than plant-based products (Audsley et al. 2009; Carlsson-Kanyama and Gonzalez 2009; Gonzalez et al. 2011). In UK, analysis of diets showed a positive relationship between amount of GHG emissions and the amount of animal products consumed in a standard 2000 kcal diet (Scarborough et al. 2014). Quantification of carbon footprints of Indian food items revealed that GHG emission from mutton is maximum (Pathak et al. 2010). The CFP of non-vegetarian meal with mutton is 1.8 times that of a vegetarian meal; a non-vegetarian meal with chicken has 1.5 times CFP of vegetarian meal while a lacto-vegetarian meal has 1.4 times. Production of red meat was found to be causing maximum GHG emission in US (Kling and Hough 2010). More than 1/3 of the GHG emission in US agricultural system comes from the livestock industry. Fish and dairy are the second most carbon intensive having 68 and 58% emissions of red meat. Cereals, oils, sugars and nuts are found to be least carbon intensive among all (Fig. 9). Over the past few decades there has been a change in food demand. The growth in income was translated into demand for higher protein nutrition (Delgado et al. 1999). In Asia, per capita meat consumption increased by a factor of 15 since 1961 and in China it increased by 130% since 1990 (FAO 2012).

Dietary shift can be a more effective means of lowering Energy and CFP of food chain (Weber and Matthews 2008). Shifting calories from red meat and dairy products to chicken, fish, eggs, or a vegetable-based diet will result in more GHG reduction

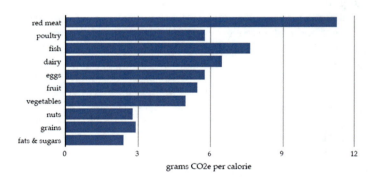

Fig. 9 Greenhouse gas emission intensities of different foods (*Source* Kling and Hough 2010)

than buying locally sourced food. Land use change from beef feed to poultry feed production can meet the caloric and protein demands of about additional 120 and 140 million people in the US (Shepon et al. 2016).

9 Food Waste

One of the major problems of food production and utilization system is the waste generated. In 2011, the FAO reported that one third of the global food production is lost as waste, corresponding to approximately 1.3 billion tons each year (Gustavsson et al. 2011; FAO 2013). Food wasted means wastage of resources, energy, time and effort in producing such product. It implies that an increased CFP per unit of food consumed. However, in most cases, avoidance of wasting food is related with the use of preserving technologies which are carbon and energy intensive. Food waste dumped in landfills produces methane during its degradation which is emitted to the atmosphere. The impact of food waste in terms of global emission is 3.3 Gtonnes of CO_2equivalent (FAO 2013). The major contributors to the CFP of food waste in a Danish Supermarket is the waste generated from food preparations followed by dairy products and meat products (Silva and Campos 2015). Food industries have the potential to improve their energy efficiency as well as re-design the logistics networks which can help in reducing the carbon footprint. Food waste can be utilized for generation of biogas. Mini biogas power plant (600 kW capacity) was first launched in Malaysia from food waste (Mydin et al. 2014). Such mini biogas plant is easy to set-up and can be transferred easily.

10 Uncertainties and Challenges

There are different uncertainties related to methodological issues, data availability and quality of data in CFP and energy input assessment of food products. GHG emission data have a high 'inherent uncertainty', because of lack of proper measurements, complexity of the process and natural variations in the system (Flysjö et al. 2009). There is a need to have uniform guidelines for quantification of CFP and energy footprints of food industries. As we move up the tiers for GHG analysis the complexity and data intensiveness increases. As per the IPCC guidelines, sources of uncertainties should be mentioned during reporting of CFP values. Emission factors and energy input factor for every process in a food industry should be available. Besides this, accounting of energy input and GHG emission for procurement of raw materials is also a great challenge. Measured data on energy consumed and GHG emission for every step may not be available, which leads the increase in uncertainties in energy foot print and CFP estimation. More studies are needed on analysis of energy foot print and CFP of food industries which can help in assessing the GHG emission from different technologies.

11 Conclusions and Future Thrust

A quick analysis indicates a wide variation in energy requirement and CFP among major food products and even within the food product types that are already known to be carbon intensive (Fig. 10). This implies that (1) there is a variation in efficiency of food processing system even for a given type of product (2) and among different types of food products with respect to the energy and CFP. . Moreover, Drewnowski et al. 2015 has reported that grains and sweets had lowest GHG emissions (per 100 g and 100 kcal) but had high energy density and a low nutrient content. On the other hand, more-nutrient-dense animal products, including meat and dairy, had higher GHG emission values per 100 g but much lower values per 100 kcal. In general, a higher nutrient density of foods was associated with higher GHGEs per 100 kcal, although the slopes of fitted lines varied for meat and dairy compared with fats and sweets.

Major inferences from such situation include (1) need to develop high energy use efficient and low CFP technologies for product development in various types of food inducts (2) identification and dissemination of high energy use efficient and low CFP protocols for each type of food product (3) changing the food habits to consumption of low Energy and CFP product (4) quantification of energy intensive and energy leakage points to reduce CFP of a specific product and (5) promotion of a product with similar source, nutrient status, taste and price but with low carbon and energy foot print. Apart from these there is also a need to provide a universal protocol for energy foot print and CFP estimation to minimize uncertainties due to estimation methods.

Food supply chains often have an impact on the environment which remains undiagnosed and unaccounted. Recently there is a growing interest in reducing the energy and CFP of food products. Some specific measures to reduce the CFP of food industries include more efficient production technologies, improving energy use efficiency,

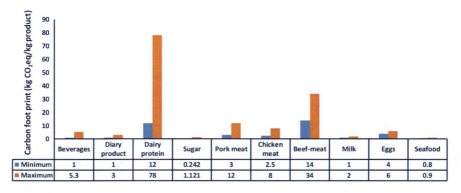

Fig. 10 Variation in CFP among major food products and food product types

development of biodegradable food packaging material, re-design of logistics network, using multimodal transport, reducing wastage of food and changing the dietary patters.

References

Amienyo D, Azapagic A (2011) Sustainability assessment of the UK beverage sector: a life cycle approach. University of Manchester, Manchester

Arroyo-Pitacua A, Márquez-Benavides L, Moreno-Goytia EL (2015) Energy requirements in a technified poultry farm in Central Mexico. In: WIT transactions on ecology and the environment, vol 195, © 2015. WIT Press https://doi.org/10.2495/esus150201

Audsley E, Brander M, Chatterton J, Murphy-Bokern D, Webster C, Williams A (2009) How low can we go? An assessment of greenhouse gas emissions from the UK food system and the scope to reduce them by 2050. Food Climate Research Network & WWF, London, UK

Barclay JMG (2012) Meat, a damaging extravagence: a response to Grumett and Gorringe. The Expository Times 123(2):70–73. https://doi.org/10.1177/0014524611418580

Baxevanou C, Fidaros D, Bartzanas T, Kittas C (2014) Energy consumption in poultries—energy audits in Greece. Int Conf Agric Eng, AgEng C0623, 1–8

Baxevanou C, Fidaros D, Bartzanas T, Kittas C (2017) Energy consumption and energy saving measures in poultry. Energy and Environ Eng 5(2):29–36

Berners-Lee M, Hoolohan C, Cammack H, Hewitt C (2012) The relative greenhouse gas impacts of realistic dietary choices. Energy Policy 43:184–190

Boopendranath MR, Shahul Hameed M (2013) Gross energy requirement in fishing operations. Fish Technol 50(2013):27–35

BIER (Beverage Industry Environmental Roundtable) 2010 Beverage industry sector guidance for greenhouse gas emission reporting. Retrieved from http://www.bieroundtable.com/files/ Beverage_Industry_Sector_Guidance_for_GHG_Emissions_Reporting_v2.0.pdf

Bokern DA (2010) Understanding the carbon footprint of our food. Complete Nutr 10(5):61–63

Canning P, Ainsley C, Huang S, Polenske KR, Waters A (2010) Energy use in the U.S. Food System. http://www.ers.usda.gov/media/136418/err94_1_.pdf

Carbon Trust (2005) Energy use in pig farming. Available online at https://pork.ahdb.org.uk/media/ 39721/energy_use_in_pig_farms_carbon_trust.pdf

Carbon Trust (2007) Industrial energy efficiency accelerator guide to the dairy sector. Available online at https://www.carbontrust.com/media/206472/ctg033-dairy-industrial-energy-efficiency. pdf

Carbon Trust (2012) Food and drink processing: Introducing energy saving opportunities for business. Available online at http://www.carbontrust.co.uk/publications/publicationdetail

Carlsson-Kanyama A, Gonzalez A (2009) Potential contributions of food consumption patterns to climate change. Am J Clin Nutr 89:1S–6S

Chhabra A, Manjunath KR, Panigrahy S, Parihar JS (2013) Greenhouse gas emissions from Indian livestock. Clim Change 117:329–344

Chakrabarti B, Kumar SN, Pathak H (2015) Carbon footprint of agricultural products. In: Muthu SS (ed) The carbon footprint handbook. CRC Press. Taylor & Francis Group, pp 431–449

Christopher W, Matthews HS (2008) Food-miles and the relative climate impacts of food choices in the United States. Environ Sci Technol 42:3508–3513

CMFRI (2014) Annual Report 2013–14. Central marine fisheries research institute, Cochin, 274 p

Čuček L, Klemeš JJ, Kravanja Z (2012) A review of footprint analysis tools for monitoring impacts on sustainability. J Clean Prod 34:9–20

Dalgaard T (2000) Farm types — how can they be used to structure, model and generalise farm data? In: Weidema BP, Meeusen MJG (eds) Agricultural data for life cycle assessments. Report 2.00.01,

Agricultural economics research institute, The Hague, The Netherlands, ISBN 90-5242-563-9, pp 98–114

de Figueiredo EB, Panosso AR, Romão R and Scala Jr NL (2010) Greenhouse gas emission associated with sugar production in southern Brazil. In: de Figueiredo et al. Carbon balance and management, vol 5, p 3. http://www.cbmjournal.com/content/5/1/3

Drewnowski A, Rehm CD, Martin A, Verger EO, Voinnesson M, Imbert P (2015) Energy and nutrient density of foods in relation to their carbon footprint. Am J Clin Nutr 101(1):184–91. https://doi.org/10.3945/ajcn.114.092486

Dunkley CS, Dunkley KD (2013) Greenhouse gas emissions from livestock and poultry. Agric, Food Anal Bacteriol 3(1):17–29

Esengun K, Erdal G, Gunduz O, Erdal H (2007) An economic analysis and energy use in stake-tomato production in Tokat province of Turkey. Renew Energy 32:1873–1881

Fang K, Heijungs R, de Snoo GR (2014) Theoretical exploration for the combination of the ecological, energy, carbon, and water footprints: overview of a footprint family. Ecol Indicators 36:508–518

FAO (2010a) Greenhouse gas emissions from the dairy sector—a life cycle assessment. In: Gerber P, Vellinga T, Opio C, Henderson B, Steinfeld H. FAO, Rome

FAO (2012). Food and agriculture organization of the United Nations. Data accessed on 30 Aug 2012 at http://faostat.fao.org/

FA (2013) Food waste footprints FAO. http://www.fao.org/fileadmin/templates/nr/sustainability_pathways/docs/Factsheet_FOODWASTAGE.pdf

FAO (2010b) Food and agriculture organization. Greenhouse gas emissions from the dairy sector: a life cycle assessment. Animal Production and Health Division, Rome

FAOSTAT (2012) FAO statistical database. Food and Agriculture Organization. http://faostat.fao.org/default.aspx

FAOSTAT (2013) FAO statistical yearbook 2013: world food and agriculture. ISSN: 2225-7373. Food and Agriculture Organization of the United Nations, Rome, 2013

Feld R, Poremski HJ (2016) Vertical guidance on an adapted energy management system (EnMS) managing energy efficiency in the beverage industry (Fruit Juice). Available online at http://www.twinningisrael.info/resources/VerticalGuidanceEnMSBeverage.pdf

Feliciano M, Rodrigues F, Gonçalves A, Santos JMRCA, Leite V (2014) Assessment of energy use and energy efficiency in two Portuguese slaughterhouses. Int Sch Sci Res Innov 8(4):253–257

Fereday N, Forber G, Powell N, Todd M, Midgely C, Nutt T, Girardello S, Wagner O (2010) Assessing the carbon footprint of cane and beet sugar production. Sugar and sweeteners, sweetener analyses. LMC International Ltd, June issue. Available online at http://www.lmc.co.uk/Articles.aspx?id=1&repID=75

Fisher J (2013) The variability and drivers of the carbon footprint of cane sugar. Int Sugar J 115(1379):782–793

Flysjö A, Cederberg C, Johannesen JD (2009) Carbon footprint and labelling of dairy products—challenges and opportunities. In: Proceeding of Joint Actions on Climate Change Conference, Aalborg, Denmark. Accessed 15 June 2013. http://www.dairyfootprint.org/research/file-cabinet

Fritzson A, Berntsson T (2006) Efficient energy use in a slaughter and meat processing plant opportunities for process integration. J Food Eng 76:594–604

Garnett T (2008) Cooking up a storm. Food, greenhouse gas emissions and our changing climate. Food Climate Research Network, Guildford

Gill M, Smith P, Wilkinson JM (2010) Mitigating climate change: the role of domestic livestock. Animal 4:323–333

Gonzalez A, Frostell B, Carlsson-Kanyama A (2011) Protein efficiency per unit energy and per unit greenhouse gas emissions: potential contribution of diet choices to climate change mitigation. Food Policy 36:562–570

Greenwich C (2010) Bottled water shown to have lightest environmental footprint among package drinks. Retrieved from http://www.nestle-watersna.com

Gustavsson J, Cederberg C, Sonesson U, Otterdijk R, Meybeck A (2011) Global food losses and food waste. FAO, Rome

Haciseferogullari H, Acaroglu M, Gezer I (2003) Determination of the energy balance of the sugar beet plant. Energy Sour 25:15–22

Hornborg S, Ziegler F (2014). Aquaculture and energy use: a desk-top study https://swemarc.gu.se/digitalAssets/1536/1536133_publication—energy-use-in-aquaculture.pdf

INCCA (Indian Network for Climate Change Assessment) (2010) Assessment of the greenhouse gas emission: 2007. The Ministry of Environment & Forests, Government of India, New Delhi

IPCC (2000) Intergovernmental panel on climate change: good practice guidance and uncertainty in national greenhouse gas inventories. Section 4: Agriculture, 94 pp. Available from: http://www.ipccnggip.iges.or.jp/public/gp/gpg-bgp.htm

IPCC (2006) Intergovernmental panel on climate change: guidelines for national greenhouse gas inventories. Agric For Other Land Use 4:87 pp

IPCC (2007) The physical science basis. In: Solomon S, Qin D, Manning M, Chen Z, Marquis M, Averyt KB, Tignor M, Miller HL (eds) (2007) Climate change contribution of working group i to the fourth assessment report of the inter-governmental panel on climate change. Cambridge University Press, Cambridge, United Kingdom and New York, NY, USA

Iribarren D, Moreira MT, Feijoo G (2010) Life cycle assessment of fresh and canned mussel processing and consumption in Galicia (NW Spain). Resour Conserv Recycl 55:106–117

ITC (2012) Product carbon footprinting standards in the agri-food sector. Technical paper. International Trade Centre, Geneva. ITC, 2012. xiii, 46 p

Janzekovic M, Mursec B, Vindis P, Cus F (2009) Energy saving in milk processing. J Achievements Mater Manufact Eng 33(2):197–203

Kalla AM, Krishna KN, Devaraju R (2017) Energy efficient and cost saving practices in dairy industries: a review. Int J Appl Eng Res Dev (IJAERD) 7(3):1–10

Karjalainen P (2013) The carbon footprint of the Finnish beverage industry for years 2000–2012 as calculated with CCaLC. Master's Thesis. Environmental Change and Policy, Faculty of Biological and Environmental Sciences, University of Helsinki

Kim D, Thoma G, Nutter D, Milani F, Ulrich R, Norrisb G (2013) Life cycle assessment of cheese and whey production in the USA. Int J Life Cycle Assess 18:1019–1035

Kitzes J, Galli A, Bagliani M, Barrett J, Dige G, Ede S, Erb K, Giljum S, Haberl H, Hails C, Jolia-Ferrier L, Jungwirth S, Lenzen M, Lewis K, Loh J, Marchettini N, Messinger H, Milne K, Moles R, Monfreda C, Moran D, Nakano K, Pyhälä A, Rees W, Simmons C, Wackernagel M, Wada Y, Walsh C, Wiedmann T (2009) A research agenda for improving national ecological footprint accounts. Ecol Econ 68:1991–2007

Khan S, Hanjra MA (2009) Footprints of water and energy inputs in food production – global perspectives. Food Pol 34:130–140

Klenk I, Landquist B, Oscar Ruiz de Imaña (2012) The product carbon footprint of EU beet sugar sugar industry 137(3):169–177

Kling MM, Hough IJ (2010) The American carbon foodprint: understanding your food's impact on climate change. Brighter Planet, Inc. http://brighterplanet.com

Lal R (2010) Managing soils for a warming earth in a food-insecure and energy-starved world. J Plant Nut Soil Sci 173:4–5

MacLeod M, Gerber P, Mottet A, Tempio G, Falcucci A, Opio C, Vellinga T, Henderson B, Steinfeld H (2013) Greenhouse gas emissions from pig and chicken supply chains—a global life cycle assessment. Food and Agriculture Organization of the United Nations (FAO), Rome

Masanet E, Worrell E, Graus D, Galitsky C (2008) Energy efficiency improvement and cost savings opportunities for the fruit and vegetable processing industry. Report by the Ernest Orlando Lawrence Berkeley National Laboratory. Publication Number LBNL-59289R

Mbohwa C (2013) Energy management in the South African sugar industry. In: Proceedings of the World Congress on Engineering 2013, vol I, WCE 2013, 3–5 July 2013, London, UK

McMichael AJ, Powles JW, Butler CD, Uauy R (2007) Food, livestock production, energy, climate change, and health. Lancet 370:1253–1263

Mendoza TC, Samson (2002) Energy costs of sugar production in the Philippine context. Philippine J Crop Sci 27(2):17–26

Mendoza TC, Samson R, Elepano AR (2004) Renewable biomass fuel as 'Green Power' alternative for sugarcane milling in the Philippines. Philippine J Crop Sci 27(3):23–39

Metz B, Davidson O, Swart R, Pan J (2001) (eds) Climate change 2001: mitigation. Contribution of working group III to the third assessment report of the intergovernmental panel on climate change (IPCC). Cambridge University Press, Cambridge and New York

Meul M, Nevens F, Reheul D, Hofman G (2007) Energy use efficiency of specialised dairy, arable and pig farms in Flanders. Ag Ecosys Environ 119:135–144

Munguia N, Velazquez L, Bustamante TP, Perez R, Winter J, Will M, Delakowitz B (2016) Energy audit in the meat processing industry—a case study in Hermosillo, Sonora Mexico. J Environ Prot 7:14–26

Mydin MAO, Nik Abllah NF, Md Sani N, Ghazali N, Zahari NF (2014) E3S web of conferences 3, 01012 https://doi.org/10.1051/e3sconf/20140301012

Okos M, Rao N, Drecher S, Rode M, Kozak J (1998) Energy usage in the food industry: a study, Report nr. IE981. American Council for an Energy-Efficient Economy

Oviedo-Rondon E (2009) Ahorro energético en granjas avícolas, (Energy savings in poultry farms, in Spanish) Proceeding of the XLVII Symposium Científico de Avicultura, Zaragoza

Padfield R, Papargyropoulou E, Preece C (2012) A preliminary assessment of Greenhouse Gas emission trends in the production and consumption of food in Malaysia. Int J Technol 1:56–66

Pandey D, Agrawal M, Pandey JS (2011) Carbon footprints: current methods of estimation. Environ Monit Assess 178:135–160

Pasqualino J, Meneses M, Castells F (2010) The carbon footprint and energy consumption of beverage packaging selection and disposal. J Food Eng 103(2011):357–365

Pathak AN (1999) Energy conservation in sugar industries. J Sci Ind Res 58:76–82

Pathak H, Jain N, Bhatia A, Patel J, Aggarwal PK (2010) Carbon footprints of Indian food items. Agric Ecosys Environ 139:66–73

Prabhakar PK, Srivastav PP, Murari K (2015) Energy consumption during manufacturing of different dairy products in a commercial dairy plant: a case study. Asian J Dairy & Food Res 34(2):98–103

Ramírez CA, Patel M, Blok K (2006) How much energy to process one pound of meat? A comparison of energy use and specific energy consumption in the meat industry of four European countries. Energy 31:2047–2063

Rees WE (1992) Ecological footprints and appropriated carrying capacity: what urban economics leaves out. Environ Urban 4:121–130

Reyes ME, Salas C, Coon CN (2011) Energy requirement for maintenance and egg production for broiler breeder hens. Int J Poultry Sci 10(12):913–920

Saxe H (2010) LCA-based comparison of the climate footprint of beer versus wine and spirits. Fødevareøkonomisk Institut, Københavns Universitet (Report; No. 207)

Scarborough P, Appleby PN, Mizdrak A, Briggs ADM, Travis RC, Bradbury KE, Key TJ (2014) Dietary greenhouse gas emissions of meat-eaters, fish-eaters, vegetarians and vegans in the UK. Clim Change 125:179–192

Setzer T (2005) Ökoeffizienz -Analyse Nachwachsende Rohstoffe zur Chemikalienherstellung am Beispiel Zucker. Masterarbeit im Studiengang Wirtschaftsingenieurwesen der Fachhochschule Mannheim

Shepon A, Eshel G, Noor E, Milo R (2016) Energy and protein feed-to-food conversion efficiencies in the US and potential food security gains from dietary changes. Environ Res Lett 11:105002

Silva P, Campos M (2015) Food waste carbon footprint: a case study at Dansk Supermarked A/S. Master's Thesis, Aalborg University, Environmental Management and Sustainability Sciences, Denmark

Sonesson U, Mattson B, Nybrant T, Ohlsson T (2004) Industrial processing versus home cooking—an environmental comparison between three ways to prepare a meal. Ambio 34(4–5):411–418

Smedman A, Lindmark Månsson H, Drewnowski A, Edman AK (2010) Nutrient density of beverages in relation to climate impact. Food and Nutr Res 54:5170

Smith DW (2014) Contribution of greenhouse gas emissions: animal agriculture in perspective. Animal agriculture & climate change. Department of Biological & Agricultural Engineering, Texas A & M University

Statistics Sweden (2002) Compiled data. Stockholm, Sweden

Steinfeld H, Gerber P, Wassenaar T, Castel V, Rosale M, de Haan C. (2006) Livestock's long shadow: environmental issues and option. FAO, Livestock, environment and development initiative Rome. FAO, Rome

Stöglehner G (2003) Ecological footprint – a tool for assessing sustainable energy supplies. J Clean Prod 11:267–277

Tan RR, Culaba AB (2009) Estimating the carbon footprint of tuna fisheries. WWWF Binary Item 17870:14

Thomassen MA, van Calker KJ, Smits MCJ, Iepema GL, de Boer IJM (2008) Life cycle assessment of conventional and organic milk production in the Netherlands. Agric Syst 96:95–107

Todde G, Murgia L, Caria M and Pazzona A (2018) A comprehensive energy analysis and related carbon footprint of dairy farms, Part 1: direct energy requirements. Energies 11:451. https://doi.org/10.3390/en11020451

Tremblay V (2000) The 1997 farm energy use survey—statistical results. Contract Report for the Office of Energy Efficiency, Natural Resources Canada. Contact David McNabb, Ottawa, ON

Troell M, Tyedmers P, Kautsky N, Ronnback P (2004) Aquaculture and energy use. In: Cleveland C (ed) Encyclopedia of energy. Elsevier, St. Louis, MO, pp 97–108

Tyedmers P (2004) Fisheries and energy use. In: Cleveland C (ed) Encyclopedia of energy, vol 2, pp 683–693

Tyedmers PH, Watson R, Pauly D (2005) Fuelling global fishing fleets. Ambio 34(8)

UNEP (United Nations Environment Programme) (2008) Planning for change: guidelines for national programmes on sustainable consumption and production.s.l. UNEP

Uzal S (2013) Comparison of the energy efficiency of dairy production farms using different housing systems. Environ Prog Sustain Energy 32(4). https://doi.org/10.1002/ep

Vergé XPC, Dyer JA, Desjardins RL, Worth D (2008) Greenhouse gas emissions from the Canadian beef industry. Agric Syst 98:126–134

Vergé XPC, Maxime D, Dyer JA, Desjardins RL, Arcand Y, Vanderzaag A (2013) Carbon footprint of Canadian dairy products: calculations and issues. J Dairy Sci 96:6091–6104

Vivekanandan E (2011) Climate change and Indian marine fisheries, central marine fisheries research institute. Special Publication, vol 105, p 97

Vivekanandan E, Singh VV, Kizhakudan JK (2013) Carbon footprint by marine fishing boats of India. Curr Sci 105(3):361–366

Vlek P, Rodraguez-Kuhl G, Sommer R (2004) Energy use and CO_2 production in tropical agriculture and means and strategies for reduction or mitigation. Environ Develop Sustain 6(1):213–233

de Vries M, de Boer IJM (2010) Comparing environmental impacts for livestock products: a review of life cycle assessments. Livestock Sci 128(2010):1–11

Wackernagel M, Onisto L, Bello P, Linares AC, López Falfán IS, Méndez García J, Guerrero AIS, Suárez Guerrero MG (1999) National natural capital accounting with the ecological footprint concept. Ecol Economics 29:375–390

Wackernagel M, Rees W (1996) Our ecological footprint. Reducing human impact on the earth. New Society Publishers, Gabriola Island, BC

Wakeland W, Cholette S, Venkat K (2012) Food transportation issues and reducing carbon footprint. In: Boye JI, Arcand Y (eds) Green technologies in food production and processing, food engineering series. https://doi.org/10.1007/978-1-4614-1587-9_9. Springer Science+Business Media, LLC, pp 211–236

Walsh C, Moles R, O'Regan B (2010) Application of an expanded sequestration estimate to the domestic energy footprint of the Republic of Ireland. Sustainability 2:2555–2572

Weber CL, Matthews HS (2008) Food-miles and the relative climate impacts of food choices in the United States. Environ Sci Technol 42(10):3508–3513

Weidema B, Thrane M, Christensen P, Schmidt J, Løkke S (2008) Carbon footprint—a catalyst for life cycle assessment. J Ind Ecol 12(1):3–6

Winther U, Ziegler F, Skontorp Hognes E, Emanuelson A, Sund V, Ellingsen H (2009) Carbon footprint and energy use in Norwegian seafood products. SINTEF

World Bank, Modi V, McDade S, Lallement D, Saghir J (2007) Energy services for the millennium development goals. Energy sector management assistance programme, United nations development programme, UN Millennium Project, and World Bank, New York

Yuttitham M, Shabbir HG, Chidthaisonga A (2011) Carbon footprint of sugar produced from sugarcane in eastern Thailand. J Cleaner Prod 19(17–18):2119–2127

Energy Footprints of Textile Products

P. Senthil Kumar and G. Janet Joshiba

Abstract The three most essential need of a human being in this world are food, clothes and shelter. On the whole, the development of textile industry will undoubtedly be tremendous, as it satisfies the second essential necessity of man. Everyone needs to hit an impression with various trendy and popular garments. This upgraded lifestyle in human beings, on the other hand, leads to many dreadful negative impacts on our environment. Energy footprint is a measure of amount of CO_2 emanated from a particular region. This approach in textile industries mainly focus on the amount of CO_2 utilized and discharged from this industry and it helps to feature the issue and make ready for restorative move to be made. The emission of Carbon dioxide into the environment in a higher amount leads to global warming and other dreadful consequences in the processing steps of textile manufacturing. The production and manufacturing sector of the textile industries mandatorily need some energy for their functioning. The amount of energy required may vary according to the processing step, apparatus and parametric conditions. Energy sources are one of the basic necessities of the textile industry in manufacturing, transporting, maintaining of the goods and it plays a major role in every other process. The energy is taken in the form of oil, gas, coal, electricity, etc. Which are all the major sources for the production of CO_2, thus estimating the energy footprints of the textile industry helps us to understand the extent of CO_2 emission and help in lowering the negative impacts of the textile industry towards our environment.

Keywords Carbon dioxide · Textile industries · Environment · Energy footprints Global warming

---

P. Senthil Kumar (✉) · G. Janet Joshiba
Department of Chemical Engineering, SSN College of Engineering, Chennai 603110, India
e-mail: senthilchem8582@gmail.com; senthilkumarp@ssn.edu.in

© Springer Nature Singapore Pte Ltd. 2019
S. S. Muthu (ed.), *Energy Footprints of the Food and Textile Sectors*, Environmental
Footprints and Eco-design of Products and Processes,
https://doi.org/10.1007/978-981-13-2956-2_3

1 Introduction

The development of science and technology has led to the much new invention on our planet earth. Since, from the beginning, Humankind has worked hard for the development of transportation and industrial sectors which plays a major role in the activities our everyday life. These anthropogenic activities such as ignition of fossil and nonfossil fuels in the transportation, manufacturing, business utilize, and other private utilizations lead to the discharge of carbon dioxide gas which occupies a major part of the dry air. Energy footprint measures the land required to ingest the carbon dioxide emitted from various anthropogenic activities in the earth. The emission of carbon dioxide on high amount causes serious ill effects to human beings and other living organisms. The above mentioned anthropogenic activities are most importantly responsible for this carbon dioxide emission (Palamutcu 2015). The consumption of energy is increasing every day as there is an increase in the needs of an individual and many developed as well as developing countries are in demand for fossil fuels. For meeting the energy demands of the human population many alternative energy sources have been discovered and used globally in several countries (Jones et al. 2015). In the most recent decade, the mandates for vitality securing, sustainable power source, vitality and upgrading of industrial science are most important factors enhancing the production of new alternative energy sources. The current pattern toward growing more household vitality sources is driven to a limited extent by political insecurity in some oil-rich countries and to some degree by the want to keep up vitality security. Simultaneously, acknowledgment of the potential social and organic repercussions of environmental climatic change is also a reason for directing emanations by growing carbon-unbiased wellsprings of energy sources such as solar energy and wind energy (Jones et al. 2015). The ecological footprint is a measure of land and water required to manage the functioning of resources and environmental distribution to fulfill the last utilization of a characterized human populace. In consideration, of the book "Our common future" released in 1987, the idea of economic advancement has come into the picture and it paved the way to various investigations related to the growth and development of the economy, Furthermore, it has also provided many new strategies for assessing the human needs. The ecological footprint is one of the methods for measuring the reliance of humankind on the environment (Ferng 2002). On the outcome of growing demand for energy, the need to evaluate the emission of carbon dioxide increases, to sort out the problem and pave the way for corrective action to be taken. Customarily, human culture has progressed without any specifications that are essential due regard to the tenets and procedures regulating the soundness of the ecosystem. The non-inexhaustible energy sources have achieved a major place in the production and manufacturing sector of industries nowadays. The energy sources have paved way for progress in the industrial science and economy of the country (Krivtsov et al. 2004).

The dress is one of the major parts of our lifestyle and it is used to cover our body, as well as to protect the body from the environment. Textile industries have always occupied a special place in the environment as it is always an important tool in redefining a human's lifestyle. The term textile encompasses all the varieties of textile products and handling steps beginning from the fiber developing and assembling to the final result preparing steps including garments manufacturing step. Every step in a textile industrial process requires energy sources to operate the machinery and equipments. Not only in the manufacturing sector but also other sectors such as transport, lighting, cooling, aerating, operating machinery, etc., everything needs the energy to drive that process. The electricity is one of the main energy source generally expended everywhere throughout the world. For another process such as heating the oil, gas and coal are used in the textile industry (Palamutcu 2015). This chapter illustrates about the impact of energy footprinting in the textile industries and its features.

2 Evolution of Textile Industry

The word textile evolved from a Latin word meaning "to weave". Initially, the garments are produced and manufactured through weaving method. Textile industries came into existence in 7000 BC and the skill of graments making is more established than metal working or earthernware making. There are confirmations that have been uncovered which has demonstrated that individuals in Harrapan human progress knew weaving and turning which remains to be the time of textiles and garments in India. Later, in the medieval period, William Lee invented a hand-worked weft sewing machine. Be that as it may, the genuine development of materials in the field of innovation occurred in the mechanical age. One name which discovers worth saying here is Sir Richard Arkwright. A visionary and doyen of his measures, he imbued the genuinely necessary innovation by then of time to give turning and weaving a mechanical standpoint through his creations (The Indian Textile Journal 2014). Later on, much new technological advancement has been made in the textile industry for producing garments. The textile industry is one of the finest fundamental businesses found globally and it is said to be a vital part of human life because it furnishes with the fundamental prerequisite called garments. This, textile industries are commonly characterized as a more impressive industry as it is upgraded every day and it determines the lifestyle of a person. The garments are made up of various kinds of fibers and other crude materials in the textile industry. Traditionally, the raw materials for the production of garments are derived by the ranchers from the animals such as cotton, sheep, silkworms, etc. Then, the fiber is synthesized by researchers through various methods using several chemical methods. The textile industry is one of the important industries in increasing the economic growth of a country globally. Due to the higher utilization of garments and apparels, the fiber generation has progressively increased globally up to 97.8 million tons by the year 2016. Even though, the textile industries are not an energy concentrated industry like the metal refining industry and smelting industries, they expend a lot of energy sources because they are basically composed of several number of plants which are fully dependent on energy for operation (Wang et al. 2017).

3 Energy Sources and Its Demand

Energy is defined as the ability and the driving force to execute a work. According to the law of conservation of energy, energy may transform from one form to another form but the total energy of a system remains constant. Energy can neither be created nor be destroyed. The discovery of fire by man for cooking and warming is the initial step of utilization of energy. For a few thousand years human energy requirements were met just by sustainable power sources such as sun, biomass, wind and water control. The energy is majorly classified into two types such as renewable and non-renewable energy sources.

Some of the commonly used units of energy are given in Table 1.

The energy appears in various forms such as Kinetic energy, Potential energy, thermal energy, Magnetic energy, Nuclear energy, Solar energy, Wind energy, Tidal energy, etc., as depicted in the (Fig. 1). Many new alternative energy sources have also been discovered by many researchers and used as a source of energy. The kinetic energy and potential energy is altogether called as mechanical energy and it is involved in the movement of objects. Energy can be transformed from one form to another form. The electricity is the highly preferred and prioritized energy source of all other energy sources.

Table 1 Commonly used units of energy

Units	Description
J	Joule
Cal	Calorie
eV	Electron volt
W	Watt
BTU	British Thermal Unit
Bbl	Barrel
Boe	Barrel of oil equivalent
hP	Horse power
kWh	Kilo watt hour
Tce	Ton of coal equivalent
Toe	Ton of equivalent
Mbd	Millions of barrels of oil per day
Gtoe	Gigatonne of oil equivalent
Mtoe	Megatonne of oil equivalent
Kcal	Kilo calorie
EJ	Exajoule
Quad	Quadrillion British thermal unit
TWyr	Tera Watt year

Energy Footprints of Textile Products

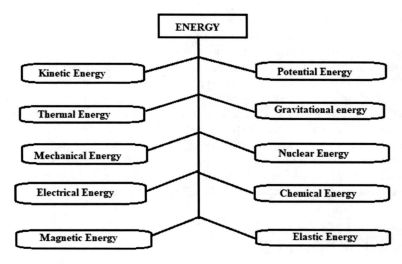

Fig. 1 Various of forms of energy

4 Energy Footprinting

Energy footprinting is the term used to clarify the expenses of environmental change due to the emission of carbon dioxide during the generation and utilization of energy sources by mankind (Palamutcu 2015). The energy footprinting measures the contribution of energy sources to meet the energy demand of a nation and it is used to indicate the consumption of energy sources globally together with its value chains (Kaltenegger et al. 2018). The energy footprint is quantified by calculating the distance of land required to absorb the emission of carbon dioxide. The aggregate size of the energy footprint in 1999 was about 6.72 billion ha. The plant earth had around 11.4 billion ha of vast space for farming and plant growth. Furthermore, the ecological footprint was around 13.65 billion ha estimated globally in 1999. The main increase in the utilization of energy sources is due to anthropogenic activities. Due to the impression of mankind, the energy footprinting expanded from 2.5 billion ha in 1961 to 6.7 billion out of 1999. The energy footprint was found to be the quickest developing part of all other footprints (GRDC). It also explains the interrelationship between emission of carbon dioxide with the consumption of energy sources by the industries. The rising urbanization rates, concentrated modern movement, environmental change, changes in land use and farming are some of the reasons which increase the emanations of carbon dioxide (Palamutcu 2015). The emission of greenhouse gases such as carbon dioxide from the textile industries is one of the major threat for environmental health as this greenhouse gases have an unfriendly effect towards nature. According to the International Energy Agency, The consumption of energy sources have expanded globally from 4661 to 9425 Mt over a period of 1973–2014. Furthermore, by the year 2014, nearly 30.5% of the world's aggregate energy sources such as gaseous petrol, oil, coal, etc., are utilized only in the industrial sector (Wang et al. 2017). Energy expenditure is directly proportional to population explosion, the increase in population shows a negative impact on energy consumption. As the population increases, the rate of urbanization gets increased, simultaneously the industrialization also plays a major role in increasing the expenditure of energy. The energy consumption of a nation fully depends on the economic growth of the country. Electricity is a main source of energy which majorly used in all sectors and nowadays it is electricity which defines a luxurious lifestyle and the industrial sectors fully depend on electricity for every processing step. Usually, the production rate of an industry is closely related with gross national product. As the production of an industry is fully depended on the electricity, so the electricity is proportional to the Gross national product (Radovic 1992).

5 Needs of Energy Footprinting

From that point onward the times of the industrial insurgency, the energy was assumed to be a fundamental part to facilitate and enhance the development of the economy. In a time period of 1990–2010 the utilization of energy sources progressively expanded to 211 EJ globally, simultaneously, the globalization has essentially modified the global economy into an intricate web of worldwide esteem chains (Kaltenegger et al. 2018). The innovations in the utilization of energy sources and vitality proficiency evaluation techniques in industry parts have expanded essentially keeping in mind the end goal to lessen the emission of carbon dioxide outflows.

6 Major Concerns in the Energy Footprinting

From the energy footprinting, it is evident that in most contextual investigations the evaluated energy footprinting sums to about portion of the aggregate figured ecological footprint also, contributes fundamentally to biological deficiencies. The extent of the vitality impression would change simultaneously when elective vitality arrangements are done. Regardless of whether energy footprinting of fossil fuel is overestimated in most contextual investigations is as of now judged by the nursery impacts confronting us today (Ferng 2002).

7 Textile Industry Manufacturing Process

The textile manufacturing is done in a sequential way and they design the garments and produce in a way that it satisfies the customer. The textile industry manufacturing process is majorly divided into four major zones such as:

Energy Footprints of Textile Products

- Spinning and yarn production
- Weaving and knitting
- Dyeing, finishing and printing
- Manufacturing of garments.

The cloth fabric is initially made from the fibre derived from the plants, silkworms, etc. The fibre is chosen with high durability, strength, fineness, luster and flexibility. The textile fibres are initially subjected to processes such as weaving, knitting, spinning, and braiding, felting and twisting. In the spinning process, the fibres are incorporated into the polymer to form continuous filaments. The fibres used for the weaving, knitting and another process usually consist of some amount of impurities and they are cleaned using the pre-treatment process. The fabric is cleaned from the dust and impurities using the processes such as desizing, scouring and bleaching. The desizing process is used to eliminate the starch materials from the fabric, whereas the scouring is used to remove the wax and fatty materials from the fabric using solvents and alkali. The bleaching process is used in removing the colour from the fabric. In the textile finishing process, the fabric is manufactured using mechanical, chemical and enzymatic treatment. Furthermore, the dyeing and printing sectors of the textile industry are the most impressive sectors of the textile which produces colourful garments of various patterns. Thus, this is the brief description of the textile manufacturing process.

8 Energy Expenditure in Textile Manufacturing

The textile business, all in all, isn't viewed as an energy concentrated industry but it involves countless operations which all together devour a lot of energy. The volume of energy consumed by a textile industry in a nation is fully depends on the structure and design of manufacturing of that textile industry, it varies from one another based on the operations involved in the textile industry. In china, around 4% of energy sources are utilized by the textile industry, whereas, in USA only 2% of energy sources are used by textile industries in U.S. (Khude 2017).

The washing, ironing and drying sectors of the textile industry are some sensitive sectors of the textile industry. Basically, the higher water and energy consumption by textile industry leads to some serious consequences such as depletion of fossil fuel, depletion of ozone layer, seasonal change and formation of photochemical. The textile business has turned out to be one of the topmost ventures among all other businesses. In this industry, there are so many ways in which energy are consumed (Wang et al. 2017). The spinning process is one of the important processes in the textile manufacturing which helps out in the formation of fabric from the fibre. This is in another way called as yarn production in which the fibre is subjected to winding, spinning, roving, etc., using the spinning equipment. The spinning process expends electricity as the main source of energy which is used to melt polymers, cooling, lighting, to run machines, etc. A major part of the energy is expended in operating the machines. In the next process, the produced yarns are weaved into clothes and garments by the process called as weaving. For an efficient production of fabric, many new technologies have been introduced in the weaving operation. Here the main expenditure of energy is in terms of electricity and heat. Most importantly electricity is required for the operation of jet looms and shuttleless looms in the weaving sector, lighting, cooling, etc. Heat is also an another important part of energy derived by the combustion of fuels such as coal, petrol, diesel and gas, then, it is required by weaving session for subjecting the fabric to desizing options. Finally, the prepared fabric are designed by impressive dyes and designing works. In the dyeing session, several types of equipments are used for the dyeing operation and it undergoes serial drying and cooling process in a cyclic way. In this dyeing section, heat is one of the main category of energy consumed for dyeing the fabrics. The heat is derived from the combustion chambers and it is produced by heating the conduction oil. The textile industry in each of its processing step, it emits an amount of carbon dioxide and other greenhouse gases (Wang et al. 2017). The fibre uptake of the textile industry have increased up to 39%, i.e. around 82 million ton in the year 2011. The textile market has a strong impact on the varying economic conditions of the country, by and large, 120,000–240,000 tons of extra fibre. Yearly fiber advertise development has been changing in the vicinity of 1.5%. As per the records of the World Trade Organization (WTO), the yearly income of the material business related with apparel is about USD 710 billion speaking to around 1.9% of world exchange volume for the year 2012. In 2015, the worldwide market for materials is a gauge to have an estimation of USD 1.6 trillion, with an increment of 32.5% since 2010. It can along these lines be underscored that the textile industry is one unavoidable part of the worldwide assembling area (Palamutcu 2015). The energy expenditure in textile industries has expanded from 12% in 2003 and reached 18-20% by current year (Khude 2017).

9 Carbon Dioxide and Its Impacts on the Environment

Researchers have found that the increasing level of carbon dioxide in the environment may lead to many serious consequences on the planet earth. The increase in the concentration of greenhouse gases and carbon dioxide in the atmosphere leads to variations in the pattern of earth's climate and also leads to global warming (Schmalensee et al. 1998). The global warming is one of the main serious consequences of the emission of carbon dioxide. The globalization, economic growth, advancement in science and technology are all some of the reasons for the emission of carbon dioxide (Narayan and Narayan 2010). There are numerous reasons that the ecological effect of the material business is such an issue, by far most of the strands delivered are engineered. It consumes a huge amount of energy sources for the production of the synthetic petrochemical-based nylon and polyester, and rayon which are some of the toxic materials contaminating the air, soil and water. The coloring and fading of textures include chemicals, vitality, and gigantic measures of water. Roughly one million tons of concoction colors are utilized each year. The wet completing procedure utilizes enormous measures of water and vitality. The textile

Energy Footprints of Textile Products

business is a massive industry and immensely leads to a lot of contamination. The material business utilizes bounteous measures of two things: water and chemicals. The manufacturing of Conventional cotton, makes up the following biggest level of overall fiber creation, is additionally intensely negative to the earth.

Fabric development requires concentrated utilization of pesticides, synthetic manures, and water. Cotton fabricating additionally requires the substantial utilization of chemicals and vitality. Wet treatment of materials like desizing, prewashing, mercerizing, coloring, printing and so on incorporates a considerable measure of concoction applications on the strands or texture. In spite of stringent environmental laws and controls, the consistency level by the material business has not been extremely palatable. In spite of the fact that, with 16% of the worldwide populace, India's offer of carbon dioxide discharges is just 3.11%, yet in one examination from the Stockholm Environment Institute it was discovered that the typified vitality of natural cotton from India was more prominent than customarily created cotton from the USA in light of the fact that the yields are considerably less in India, requiring more land to develop a similar sum, and quite a bit of India's energy source is produced by coal (Mehta and Goyal 2015). Researchers perceive the connection between a worldwide temperature alteration and environmental change. Climatic variation is one of the dreadful effects of the emission of carbon dioxide to the environment and this changes in Environment has gotten much consideration at global discussions among government officials and business pioneers in the previous decade. Due to anthropogenic activities, the level of carbon dioxide has increased leading to seasonal variation and ozone depletion. Some of the anthropogenic activities which are responsible for the emission of carbon dioxide is depicted in Fig. 2. The carbon footprinting is used to measure the emission of carbon dioxide from various sources due to human activities (Wiedmann 2009). It quantifies the emanation of gases that add to warming the planet in carbon dioxide (CO_2) reciprocals per unit of time or item. Governments, arrangement creators and organizations take the initiative to alleviate a dangerous atmospheric deviation and to look for approaches to decrease the emission of greenhouse gases outflows in light of developing interests and worries about the environmental change in the course of recent decades (Alhorr et al. 2014).

10 Types of Energy Used During Processing

They are various types of energy are involved in the manufacturing of textile goods and the amount of the energy expended for that processing step is based on the process. The most commonly used energy sources are heat, electricity, etc. (Palamutcu 2015). The textile industries utilize extensive amounts of both fuels and energize. The offer of power and powers inside the aggregate last vitality utilization of any one nation's material segment relies upon the structure of the material industry in that nation (Khude 2017). The electricity is the topmost required energy source in all type of industry and in textile industry, all major and minor process depends on electricity for operation. The textile processing steps such as weaving, knitting, spinning, and

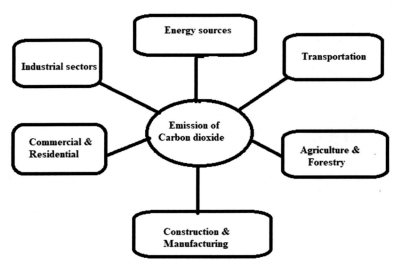

Fig. 2 Various sources of emission of carbon dioxide

all other steps require electricity for operating the machinery and equipments. Heat energy is another one important energy source used during the manufacturing of the textile products. It is used in the heating process and also in the production of steam. The dyeing sector of the textile industry needs heat energy for processing. It is evaluated from the UNIDO report of 1992, that power utilization rates in the aggregate expended vitality for singular material generation stages are, 85% for weaving, 93% for turning, 43% for wet preparing, and 65% for dress producing (Palamutcu 2015). Whatever remains of the vitality that is utilized as a part of the material handling plants is provided by other vitality wellsprings of fuel, flammable gas, and coal. Electric energy is seriously utilized for machine driving, cooling, and ventilation in the man-made fiber preparing, turning, weaving-sewing nonwoven, what's more, dress creation stages. Other than machine driving purposed utilization of electric vitality, cooling, warming, lighting, and different random utilization of power in the material plants are experienced (Khude 2017). The consumption of energy in Wet-handling processes are significant and it uses more amount of both steam and warmth. The vitality utilized as a part of wet processing relies upon different factors such as the type of the item being handled like fiber, yarn, texture, material, etc. (Hasanbeigi 2010). Not only the production sector consumes electricity all other systems such as fans, pumps, lighting and all other minor operations also requires electricity for operation. These industries also face wastage of a part of energy during some processing steps. The energy loss is higher in the motor driven systems than the other operations and the motor driven operations such as pumps, fans, etc., are the major source of wastage of energy. The spinning sector is another large entity which is a major reason behind high energy consumption of textile industries (Khude 2017).

11 Energy Footprinting Tools

Measurement of the amount of energy consumed is tracked using the energy footprinting and at the same time the amount of carbon dioxide emitted during usage of the energy sources for textile processing are determined using the energy footprinting tools. The energy footprinting is nothing but monitoring and recording the energy consumption of textile industry on a monthly basis. It is purely a documentation process. The date is gathered and stored in excel sheets for comparison and the monthly energy bills are mainly required to know the amount of energy consumed, furthermore it helps in tracking the increase or decrease in consumption of energy sources. All types of energy sources used in the manufacturing process are taken into account for drafting the energy footprint framework. According to the U.S. Department of Energy, the energy consumption is defined as the amount of energy applied in the process of manufacturing. Several types of fuels and energy are applied in the industries and the number of fuels such as coal, gas, petrol, diesel, etc., for activating the energy sources are also taken into account (U.S. Department of Energy).

Some of the key terminologies used in the energy footprinting and its description are given below in the Table 2.

12 Energy Footprinting of Textile Industry

The textile industry is one of the tremendous industry and it is also one of the mass producers of toxic wastes to the environment. Water is utilized at each phase in texture production to disintegrate chemicals to be utilized as a part of one stage, at that point to wash and flush out those same chemicals to be prepared for the subsequent stage. A huge amount of water and harmful chemicals are used in the textile process and it eliminates the wastes in a plentiful amount. In the fabric production some of the processing steps such as desizing, scouring, bleaching, prewashing, dyeing and so forth incorporates a considerable measure of synthetic chemical applications on the

Table 2 Key terms of energy footprinting (*Source* U.S. Department of Energy)

Key terms	Description
Energy consumption	The quantity of energy applied in the process
Energy source	The type of energy consumed such as electricity, natural gas, fuel oil, etc.
Relevant variables	Variables that support the energy consumption and use such as production, operating hours, heating degree days, etc.
Energy use	Application of energy
Load factor	Average load relative to the full load of the equipment
Duty factor	The average ratio of run time to operating hours
EnPI	Energy Performance Indicator

strands or texture. Amid the procedure of wet treatment, an immense amount of non-renewable energy sources is expended which have carbon content and respond with oxygen to frame carbon dioxide. This outcome in fermentation, petroleum product exhaustion and eventually a worldwide temperature alteration. Nearly, 1074 billion KWh of power are consumed by the various processing steps of the textile industry and this amount of consumption of energy sources leads to the emission of a huge amount of carbon dioxide. During the dyeing process, the chlorine is added to neutralize the colour of the garments and it leads to the emission of organochlorine which are perilous in nature. In excess of 1 million tones of material are discarded every year out of which 50 for each penny are recyclable. These wastages require a landfill, and they don't deteriorate rapidly. Woolen articles of clothing while at the same time deteriorating produce gases like methane which brings about a worldwide temperature alteration. Nearly 75% of the total carbon dioxide is used in the manufacturing and washing of the clothes and garments (Jain 2017).

13 Standards Used for Conserving Energy

The manufacturing of textile industry is a tedious process with many complicated types of machinery and it is made up of numerous unit operations. Energy is the basic component of textile industry based on which the manufacturing process takes place. Many standards and labels are used to characterize and portray the utilization of energy sources. Implementing this standards, methods and approaches in the textile industry can decrease the harmful effects caused by the industry to the environment. Controls, reports, logical works, and learns about vitality utilization estimations, natural effects assessment, and life cycle evaluation (LCA) works for textile items and generation techniques have as of late expanded as an aftereffect of wild vitality cost approaches and raising natural mindfulness around the world. Techniques, models, methodologies, and principles that are used to assign vitality impressions in the material business are outlined beneath (Palamutcu 2015).

14 Energy Conservation Act

The energy conservation act was enunciated by the Indian government in the year 2002 for conserving the energy sources to meet the future demands. This act specify energy consumption standards for notified equipment and appliances. This act prohibits manufacture, sale, purchase and import of notified equipment and appliances not conforming to energy consumption standards. It notifies energy-intensive industries, other establishments, and commercial buildings as designated consumers and also establish and prescribe energy consumption norms and standards for designated consumers. This act is helpful in implementing the conservation of energy sources (Ministry of Power).

15 Exergy Analysis

Exergy analysis is a basic apparatus for framework outline, examination and streamlining of mechanical frameworks. The exergy technique can help to achieve the objective of more proficient energy utilization. It also resolves the squanders and misfortunes found in the energy consumption of an industrial process. In addition, exergy investigation can uncover whether the energy consumption is in a right way or not (Hajjaji et al. 2012). This analysis is based on the second law of thermodynamics and mass and energy balance. The misconception in the energy expenditure is determined using the exergy analysis and it aids in giving solutions and alternatives to reduce the energy loss taking place in textile industries (Palamutcu 2015).

16 Sankey Diagrams

Sankey diagrams is an drawing tool used to picturize the inlet and outlet pathways of the steam and air. This sankey diagrams are mainly used in reduction of leakages and it helps in easy understanding of the usage and destination of the flow. In the sankey diagram the amount of the flow is denoted but the width of the arrow and the direction of the arrow denotes the direction of the flow (Palamutcu 2015).

17 Measures to Reduce Energy Wastage

The evaluation of the consumption of energy is mandatory to know the volume of energy wasted during the manufacturing of textile industry. The conservation and management of energy sources should be made mandatory to decrease the energy wastage and to lessen the emission of carbon dioxide into the atmosphere. This examination of energy source utilization in the textile business helps in process improvement and substitution of the current hardware with cutting edge new innovation (Khude 2017).

We can adopt some of the strategies to reduce the energy footprint in the textile industry and it is elaborated below in detail:

18 Efficient Electricity Management

The material business utilizes an immense number of machineries which consumes electricity as an source of energy. Traditionally, solitary motors are used where the mechanical power is created and transmitted to different parts of the machine in an aggregate way, but nowadays numerous cutting edge machines use various engines with a control board controlling the development of each engine, which is straight-forwardly coupled to a machine part to drive it autonomously from others consuming a less amount of energy. Lighting is an very important sector in the textile industry and fundamental for operation of all the sectors in textile industry. It is essential to rethink whether the light source is used in the most productive way and take power sparing measures. In textile industry, heating is an another vast sector used for steam production, gas warming, etc. Effective selection of fuel in the textile industry can decrease the high energy expenditure. Implementation of advance boilers such as Scotch type, Lancastrian boilers can improve the efficiency in manufacturing of tex-tile industries. Arrangement of pipelines in an effective way can reduce the spillage of steam and air over pipes (Khude 2017).

19 Selection of Efficient Fuel

The fuel plays an important role in the manufacturing process of textile industries and it is mandatory select an efficient fuel for the processing of textile industry. Fuel makes a process costly and it also leads to higher emission of carbon dioxide to the environment. The processes such as production of compressed air, warming, pumping, cooling, etc., are responsible for high consumption of fuel sources in textile industry. As the demand for the fuel gets increased every day the cost of the textile manufacturing also increased. Selection of alternative efficient fuel can decrease the cost of the fuel expenditure. Alternative cheaper source such as lignite is used as a fuel in the textile industries. Evaluating the proficiency of other energy sources in the textile industry can help in selection of an efficient fuel for textile processing.

20 Detecting Energy Loss

The main step in conserving energy is primarily reduction of energy loss and the chances of reducing the probability of energy loss. Recovery of the energy loss is also a clever decision in the energy management. Evaluating the chances of energy loss in the textile industry helps in designing an efficient and less energy consuming manufacturing process. Recovering the used energy from the process and again using it an energy source lower the volume of energy sources used. Monitoring the leakages and holes in the pipelines during the passage of air, steam and water should be made mandatory to conserve the energy sources. The pipelines used in the textile industries should be sealed well and insulated well to prevent leakages (Ozturk 2005).

21 Reduce, Reuse and Recycle

This concept is one of the famous approach used in industries to lessen the harmful impacts of the environment. The energy footprint can be reduced by decreasing the usage of energy and lessening the emission of carbon dioxide. The recycling of water used in the dyeing and other sectors of the textile industry can help in reducing the emission of carbon dioxide. This reduction of carbon dioxide to the environment can automatically reduce the variations caused by the environment's climate. New innovative and conservation technologies should be followed in the textile industries to reduce the negative impacts of the textile industry to the environment. Wet processing techniques followed in textile industries such as desizing, scouring, bleaching, dyeing and washing are one of the major chemical using process in the textile industry. Reducing the chemicals used in these wet process industry can reduce the emission of carbon dioxide. Implementing eco-friendly bleaching and enzyme technology in dyeing is a good alternative. Recycling of the water used in dyeing and other process helps in reducing the wastage of fresh water used in textile industry. Various types of garments can be made out of these recycled fibres and usage of the virgin fibres decreases the rate of the energy consumption in textile industry by 84% and decreases the emission of carbon dioxide up to 77% (Jain 2017).

22 Education

Awareness education is one of the remedies to decrease the energy footprint. Educating the people about the energy demands and the harmful effects of the emission of carbon dioxide helps in decreasing its effects on the environment. Proper education about the new innovative and eco-friendly techniques that are followed in other countries can be taught and implemented in our textile industry can make them aware of this textile technology technically.

23 Future Trends

Energy has been an important wealth found in nature. The electricity and heat energy are two important energy sources required by the textile industry. The conservation of energy sources and fuel sources are essential to meet the growing demand for energy sources. Implementing the reduce, reuse and recycle concept in the textile manufacturing may enhance the manufacturing process in an efficient way. Producing electricity using renewable energy may help on the conservation of energy. Developing technologies to save and store the solar energy in an efficient way may help in producing electricity in a cost-effective way. The replacement of chemical substances used in dyeing units with eco-friendly substances can amount of carbon dioxide emitted from the industry. Planning implementing executing the manufacturing process with lesser toxic substances can decrease the energy footprint.

24 Conclusion

The energy footprint of the textile industry helps in tracking the consumption of energy sources used by the textile industry for manufacturing operations. It is used in several industries to map the energy requirement of the industries. The enunciation of new approaches, standards and methodologies help in decreasing the energy footprint of the textile industry. The manufacturing, processing and packaging sector of textile industry are all energy sensitive sector and due to the usage of the high amount of energy sources the carbon dioxide is emitted into the environment. As there is a lesser amount of empty lands to sequester this carbon dioxide in our country, it leads to ozone depletion and global warming. Implementing the usage of eco-friendly chemicals in the textile industry will decrease the amount of carbon dioxide emission and other greenhouse gases. The energy footprinting is an effective tool in monitoring the industrial activities of the textile industry. This helps in understanding the power requirements of every sector of the textile industry.

References

Alhorr Y, Eliskandarani E, Elsarrag E (2014) Approaches to reducing carbon dioxide emissions in the built environment: low carbon cities. Int J Sustain Built Environ 3:167–178

Energy Footprinting, https://www.gdrc.org/uem/footprints/energy-footprint.html

Energy Footprinting Tool, Energy efficiency and renewable energy. U.S. Department of Energy, https://www.energy.gov/sites/prod/files/2017/11/f39/EnergyFootprintGuide.pdf

Evolution of textile industry in India. Indian Text J, October 2014

Ferng JJ (2002) Analysis toward a scenario analysis framework for energy footprints. Ecol Econ 40:53–69

Hajjaji N, Pons MN, Houas A, Renaudin V (2012) Exergy analysis: an efficient tool for understanding and improving hydrogen production via the steam methane reforming process. Energy Policy 42:392–399

Hasanbeigi A (2010) Energy-efficiency improvement opportunities for the textile industry. Ernest Orlando Lawrence Berkeley National Laboratory

International Energy Outlook (2017). https://www.eia.gov/outlooks/ieo/

ISO 14064 (2006) International Organization for Standardization, Geneva, Switzerland

Jain M (2017) Ecological approach to reduce carbon footprint of textile industry. Int J Appl Home Sci 4:623–633

Jones NF, Joseph P, Kiesecker M (2015) The energy footprint: how oil, natural gas, and wind energy affect land for biodiversity and the flow of ecosystem services. Bioscience 65:290–301

Kaltenegger O, Loschel A, Pothen F (2018) The effect of globalisation on energy footprints: disentangling the links of global value chains. Energy Econ 68:148–168

Khude P (2017) A review on energy management in textile industry. Innovative Energy Res 6

Krivtsov V, Wäger PA, Dacombe P, Gilgen PW, Heaven S, Hilty LM, Banks CJ (2004) Analysis of energy footprints associated with recycling of glass and plastic—case studies for industrial ecology. Ecol Model 174:175–189

Mehta R, Goyal C (2015) Role of carbon footprint in textile and apparel industry. Textile Value Chain, http://www.textilevaluechain.com/index.php/article/technical/item/235-role-of-carbon-footprint-in-textile-and-apparel-industry

Ministry of Power, Energy Conservation Act (2001), http://www.indiacode.nic.in/bitstream/123456789/2003/1/200152.pdf

Narayan PK, Narayan S (2010) Carbon dioxide emissions and economic growth: panel data evidence from developing countries. Energy Policy 38:661–666

Ozturk HK (2005) Energy usage and cost in textile industry: a case study for Turkey. Energy 20:1–23

Palamutcu S (2015) Energy footprints in the textile industry. Handbook of Life Cycle Assessment (LCA) of textiles and clothing. A volume in Woodhead Publishing Series in Textiles, pp. 31–61

Radovic LR (1992) Energy supply and demand. Energy and Fuels in Society, Available at https://www.ems.psu.edu/~radovic/Chapter5.pdf

Schmalensee R, Stoker TM, Judson RA (1998) World carbon dioxide emissions: 1950–2050. Rev Econ Stat 80:15–27

Wang L, Li Y, He W (2017) The energy footprint of China's textile industry: perspectives from decoupling and decomposition analysis. Energies 10:1461

Wiedmann T (2009) A first empirical comparison of energy footprints embodied in trade—MRIO versus PLUM. Ecol Econ 68:1975–1990